MAY - - 2009

About Island Press

Since 1984, the nonprofit Island Press has been stimulating, shaping, and communicating the ideas that are essential for solving environmental problems worldwide. With more than 800 titles in print and some 40 new releases each year, we are the nation's leading publisher on environmental issues. We identify innovative thinkers and emerging trends in the environmental field. We work with world-renowned experts and authors to develop cross-disciplinary solutions to environmental challenges.

Island Press designs and implements coordinated book publication campaigns in order to communicate our critical messages in print, in person, and online using the latest technologies, programs, and the media. Our goal: to reach targeted audiences—scientists, policymakers, environmental advocates, the media, and concerned citizens—who can and will take action to protect the plants and animals that enrich our world, the ecosystems we need to survive, the water we drink, and the air we breathe.

Island Press gratefully acknowledges the support of its work by the Agua Fund, Inc., Annenberg Foundation, The Christensen Fund, The Nathan Cummings Foundation, The Geraldine R. Dodge Foundation, Doris Duke Charitable Foundation, The Educational Foundation of America, Betsy and Jesse Fink Foundation, The William and Flora Hewlett Foundation, The Kendeda Fund, The Forrest and Frances Lattner Foundation, The Andrew W. Mellon Foundation, The Curtis and Edith Munson Foundation, Oak Foundation, The Overbrook Foundation, the David and Lucile Packard Foundation, The Summit Fund of Washington, Trust for Architectural Easements, Wallace Global Fund, The Winslow Foundation, and other generous donors.

The opinions expressed in this book are those of the author(s) and do not necessarily reflect the views of our donors.

Science
MAGAZINE'S
State of the Planet

Science
MAGAZINE'S

State
of the
Planet

2008–2009

Edited by
Donald Kennedy
and the editors of *Science* magazine

 AMERICAN ASSOCIATION FOR
THE ADVANCEMENT OF SCIENCE

 **ISLAND**PRESS
Washington • Covelo • London

ISSN 1559-1158

Library of Congress Cataloging-in-Publication
data is available on file. British Cataloguing-in-
Publication data is also available.

⊕ Printed on recycled, acid-free paper

Design by BookMatters

Manufactured in the United States of America

10 9 8 7 6 5 4 3 2 1

Contents

PROJECTING THE FUTURE

Acknowledgments

Four editors of Science—Brooks Hanson, Andrew Sugden, Caroline Ash, and Jesse Smith—were centrally engaged with planning and executing the Special Issues that formed the basis for this book. The *Science* copyediting department handled the additional revisions for this book, and Pat Fisher helped in the preparation throughout. Our colleagues in News, especially Colin Norman, helped to identify News stories to lend illumination to each chapter and add current policy content. Alan Leshner, the Chief Executive Officer of the American Association for the Advancement of Science, along with its Board of Directors, lent enthusiastic support to this project, fulfilling the AAAS mission to support science and serve society. Todd Baldwin and his colleagues at Island Press provided splendid editorial help. Finally, the authors of all these chapters are members of a scientific community in which volunteer efforts to explain research and its relationship to human welfare are among its most admirable characteristics. We thank them for their willingness to contribute in this way.

Introduction

DONALD KENNEDY

In the volume published in 2006 by Island Press, *Science*'s editors broadly evaluated the "State of the Planet," which became the name of the book. It contained authoritative accounts, authored by leading experts, that chronicled a variety of the challenges Earth faced because of the ways in which its human inhabitants were stressing its slowly renewable or nonrenewable resources. In 2007, the Annual Meeting of the American Association for the Advancement of Science (AAAS)—the professional society that publishes *Science*—was held in San Francisco on the theme of sustainability.

That meeting came at a significant turning point in time. The scientific case on climate change had already reached the critical level of density that often marks a migration from the research journals to the popular press. The cover of *Time* surprised many observers with a photograph of a distressed polar bear and the legend "BE WORRIED. BE VERY WORRIED" about climate change. Katrina had just happened, and Al Gore's film *An Inconvenient Truth* was beginning a run toward public excitement and interest that would eventually win it an Academy Award. The choice of "Sustainability" as the working theme of the meeting was significant: It underscored the point that action about climate change is an intergenerational problem. There are different versions of sustainability, but all of them relate the welfare of those alive now to that of those who will succeed them. Most of us feel a fairly serious obligation to leave our successors at least as well off (in terms of natural resources and other human needs) as we have been.

This theme is also an appropriate key to the transition between the first volume and this. Increasingly, the challenge confronting scientists concerned about the global environment now focuses on two closely related themes: energy and climate change. In this relationship, energy takes the center of the stage for a variety of reasons. Not only do many developed countries, including the United States, depend heavily on the energy supplied by the combustion of fossil fuels—coal, oil, and natural gas—but also, because nations and regions differ widely in their natural endowments of fossil fuels, situations arise in which resource-rich nations exert some forms of control over world prices, while the less well-endowed work to achieve a kind of "energy independence."

But of course the connection depends on another feature of fossil fuels. They are carbon-rich, and when they are combusted, they add carbon dioxide (CO_2) to the Earth's atmosphere. The connection between atmospheric CO_2 and our planet's heat budget had been known for centuries. CO_2 serves to conserve heat, because it and other diatomic molecules in the atmosphere are transparent to incoming radiation but block the reflected long wavelengths moving away from Earth to space. That accounts for the "greenhouse" metaphor—the "greenhouse gases" are like the panes of glass in a real greenhouse, and they prevent the temperature at Earth's surface from reaching average levels nearly 69 degrees colder than we experience now.

In his Presidential Address at the 2007 Annual Meeting, published in full in *Science* and used here as the introductory essay on the general problem of sustainability, John Holdren, the Teresa and John Heinz Professor of Environmental Policy at Harvard and Director of the Woods Hole Research Center, explores those twin links between energy and climate. Plainly, rising affluence around the world and population growth will combine to continue the current dramatic increase in the CO_2 content of the Earth's atmosphere—now at about 385 parts per million by volume (ppmv), compared with its preindustrial level of about 280 ppmv. Holdren points out that although the efficiency of the world energy economy is improving at 1% per year, that is not nearly enough to offset the current rate of increase in demand. We would have to improve efficiency much faster, and drive toward technological solutions that could reduce the demand for fossil fuels—including solar, biofuels, and conceivably nuclear. Holdren's address forms an appropriate chapter-length introduction to this volume dedicated to energy, climate, and sustainability.

Energy Solutions

Introduction

DONALD KENNEDY

I t may be helpful to begin this section with an account of various policy options regarding global warming that either have been proposed as solutions or are likely to be in the future. One is to continue our present dependence on fossil fuels while developing technologies to store or sequester emitted carbon. Another is to develop other fuel sources that are less carbon-intensive and therefore more sustainable.

Much of this opening section explores the second of these approaches, looking at several of the possible alternatives to present energy sources that might put us on a different, less carbon-intensive and therefore more sustainable path. Janez Potočnik, the European Science and Research Commissioner, sets out a program through which the European Union (EU) might achieve a dramatic reduction in the growth of carbon emissions, through a "carbon constraint" policy emphasizing renewables as sources and efficiency improvements in both buildings and the transportation sector. The program he describes is ambitious, but achieving its 2020 goal of providing about a third of EU electricity from renewables will require strong commitments from a number of nations with different needs and capacities. Potočnik expresses concern about the level of support for needed new technologies in EU's national budgets and urges participation in a regional Strategic Energy Technology Plan.

Biofuels have assumed an important role in the search for renewable sources because the world's transportation sector has been a major contributor to atmospheric CO_2, and also because the United States and many other nations would like to reduce their dependence on oil from the Middle East. The topic has become controversial, partly because of the substantial commitments made by the United States to grow corn dedicated to producing ethanol as a gasoline substitute. Many argue that this linkage of the agriculture and energy economies will raise world food prices and harm poor people and nations.

José Goldemberg, the Secretary for the Environment in São Paulo, Brazil, characterizes the role that "renewable" energy sources currently play in the global energy equation. After that exploration, he turns to report the remarkable progress of the Brazilian program to produce ethanol from sugarcane—which now accounts for over a quarter of the energy needs for their transportation sector. The economic "learning curve" suggests that a similar strategy might well work elsewhere. Chris Somerville's editorial suggests some of the difficulties that ethanol production would likely encounter in the United States, and my own sharply critical editorial, published later, points to some of the problems that derive from linking the energy and agricultural economies together in this way.

A different perspective on the biofuels question comes from David Tilman and his colleagues, a distinguished group of ecologists at the University of Minnesota who have had long experience analyzing productivity and diversity on lands of various agricultural quality. In their Report late in 2006, they undertake a comparative analysis of energy yields from mixed grassland plots on marginal soils, concluding that their low-input, high-diversity systems have higher bioenergy yields than monocultures—for example, those used to produce ethanol from corn. Advocates for other bioenergy-yielding crops like the perennial switchgrass have argued that farm-scale planting of such varieties can produce higher yields still. It seems fair to predict continuing controversy over the discussion of which systems for converting plants for renewable bioenergy will work best. At the moment, in terms of efficiency and energy yield, ethanol from sugarcane looks better than most alternatives.

Solar energy has had a long history of consideration as a renewable, nonpolluting energy source; indeed, in the early days of the Carter administration, a government-supported Solar Energy Research Institute was established in Colorado. After some progress, the effort was discontinued. But today's high oil prices have changed the economics of energy production radically and make various renewable energy solutions more cost-competitive. In "Can the Upstarts Top Silicon," Robert F. Service reports on nascent technologies that could make solar cells more efficient and less expensive. Researchers are investigating new materials, such as metal nanoparticles, that could increase the cells' thermodynamic efficiency: how much electrical energy will be produced in response to a given number of impinging photons. Further research aims to reduce costs through better manufacturing technologies. The proposed solutions are highly complex, are difficult to realize outside of the lab, and will require continued research investments. But with world oil prices in the neighborhood of $130 per barrel, and threatening to rise even higher, solar energy might be able to "back in" to a competitive position even without substantial technological improvements.

What this section suggests is that renewable, carbon-free energy systems are prepared to shoulder a substantial part of the world's energy burden. Major research breakthroughs, however, would be required for a dramatic shift in this direction: efficiency improvements in photovoltaic systems, for example, or solutions that would allow the direct use of cellulosic digestion of plant materials.

Of course, a carbon-free source of energy now contributes significantly, in the form of nuclear power. Concerns about the vulnerability of fissile materials, and worries about the disposition of hazardous nuclear waste, have made this a politically unacceptable solution in most but not all countries. News articles from *Science* by Daniel Clery and John Bohannon illustrate different novel solutions to the "renewables" problem: one involving new photovoltaic applications and another using a different kind of source material for providing nuclear power.

The combination of fossil fuel costs and environmental liabilities has encouraged all these efforts to develop and improve alternatives. Much longer-range research efforts have focused on nuclear fusion (in contrast to fission), but these appear to be many years away from practical solution; the managers of energy utilities are, as a group, highly skeptical about whether the kinds of high-end fusion reactors that have been tried could be used in practice. Other efforts claim progress toward a "hydrogen economy," in which energy is obtained in ways that do not rely on breaking carbon bonds. But most efforts in this direction, like the "hydrogen car," also require fossil fuel participation—and the idea of an all-hydrogen economy is a concept more popular as a political hope than as a reliable solution.

In short, there is hope for renewable energy sources. But much depends on new technology and on economics. A major breakthrough in the biofuels area would require a new source for cellulose that could both spare greenhouse gas emissions (by not requiring cropland conversion) and add value through displacement of petroleum-based fuels. Major work is in progress in this area, as it is in the development of improved solar systems. But the economics really matter: low petroleum prices discourage the development of competing alternatives, whereas high prices may permit more widespread adoption of renewables.

The Energy-Economy-Environment Dilemma

JOHN P. HOLDREN

In my 2007 Presidential Address to the 2007 AAAS Annual Meeting, I began with the proposition that human well-being rests on a foundation of three pillars, all of which society must take pains to preserve and enhance (1). The three are

- Economic conditions and processes, such as production, employment, income, wealth, markets, trade, and the technologies that facilitate all of these;

- Sociopolitical conditions and processes, such as national and personal security, liberty, justice, the rule of law, education, health care, the pursuit of science and the arts, and other aspects of civil society and culture; and

- Environmental conditions and processes, including our planet's air, water, soils, min-

eral resources, biota, and climate, and all of the natural and anthropogenic processes that affect them.

Arguments about which of the three pillars is "most important" are pointless, in part because each of the three is indispensable: just as a three-legged stool falls down if any leg fails, so is human well-being dependent on the integrity of all three pillars.

The issue of society's energy supply affects economy, environment, and civil society alike, but a particular dilemma resides in the countervailing roles of energy in the economy and in the environment (2–4): reliable and affordable energy is essential for meeting basic human needs and fueling economic growth, but the harvesting, transport, processing, and conversion of energy using the resources and technologies relied upon today cause a large share of the most difficult and damaging environmental problems society faces.

Contemporary technologies of energy sup-

This article first appeared in *Science* (25 January 2008: Vol. 319, no. 5862). It has been revised for this edition.

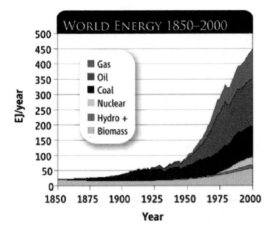

WORLD ENERGY 1850–2000

Legend: Gas, Oil, Coal, Nuclear, Hydro +, Biomass

Y-axis: EJ/year (0, 50, 100, 150, 200, 250, 300, 350, 400, 450, 500)
X-axis: Year (1850, 1875, 1900, 1925, 1950, 1975, 2000)

TABLE 1. World energy supply in 2005

	World	USA	China
Primary energy[a]	514	60	350
of which Oil	34%	40%	18%
Natural Gas	21	24	2
Coal	26	25	62
Nuclear energy	6	8	0.6
Hydropower	2	3	15
Biomass and other	11	3	15
Primary energy[b]	17,300	4,000	2,400
of which Coal	40%	50%	80%
Oil and gas	26	21	3
Nuclear	16	20	3
Hydropower	16	7	15
Wind, geothermal, and solar	2	2	0.1

[a]Exajoules. [b]Terawatt-hours.

FIGURE 1: World supply of primary energy 1850–2000 (6). Primary energy refers to energy forms found in nature (such as fuelwood, crude petroleum, and coal), as opposed to secondary forms (such as charcoal, gasoline, and electricity) produced from the primary ones using technology. "Hydro +" includes hydropower, geothermal, wind, and solar. Fossil fuels are counted at higher heating value, and hydropower is counted as energy content, not fossil fuel equivalent. One exajoule (EJ) = 10^18 joules = 0.95 quadrillion Btu.

About a third of the primary energy is devoted to electricity generation. Net electricity = gross generation less the electricity used within the generating facility. In the "primary energy" column, hydropower is counted as energy content, not fossil fuel equivalent. "Other" includes wind, geothermal, and solar energy (8).

ply are responsible for most indoor and outdoor air pollution exposure, most acid precipitation, most radioactive wastes, much of the hydrocarbon and trace-metal pollution of soil and groundwater, nearly all of the oil added by humans to the oceans, and most of the human-caused emissions of greenhouse gases that are altering the global climate (5).

The study of these environmental impacts of energy has been a major preoccupation of mine for nearly four decades. I have concluded from this effort that energy is the hardest part of the environment problem; environment is the hardest part of the energy problem; and resolving the energy-economy-environment dilemma is the hardest part of the challenge of sustainable well-being for industrial and developing countries alike.

Figure 1 shows the composition of world primary energy supply during the bulk of the fossil fuel era to date, from 1850 to 2000 (6). Energy use increased 20-fold over this period—that number being the product of a somewhat greater than fivefold increase in world population and a somewhat less than fourfold increase in average energy use per person (7). Fossil fuel use increased more than 150-fold, rising from 12% of the modest energy use of 1850 to 79% of 2000's much larger total. By 2005, fossil fuels were contributing 81% of the world primary energy supply, 82% in China, and 88% in the United States (8); even in the electricity sector (where nuclear power, hydropower, and wind, solar, and geothermal energies make their largest contributions), fossil fuels accounted for two-thirds of global generation (Table 1).

The huge increase in fossil fuel use over the past century and a half played a large role in expanding the impact of humankind as a global biogeochemical force (9), not only through the associated emissions of carbon dioxide (CO_2), oxides of sulfur and nitrogen, trace metals, and more, but also through the mobilization of other materials, production of fertilizer, transport of water, and transformations of land that the availability of this energy made possible (10). At the end of the 20th century and the beginning of the 21st, the fossil fuel–dominated energy supply system continued to impose immense environmental burdens at local, regional, and global scales, despite large investments and some success in reducing emissions to air and water per unit of energy supplied (11).

Fine particles appear to be the most toxic of the usual air pollutants resulting from the combustion of fossil and biomass fuels, and whether emitted directly or formed in the atmosphere from gaseous precursors, they have proven difficult to control (12). The concentrations of fine particulates in urban air in the United States, Western Europe, and Japan have mostly been falling in recent years, but in cities across the developing world, the concentrations have risen to shockingly high levels—often several times the World Health Organization (WHO) guidelines (11). Population exposures to particulate matter from the combustion of fossil and biomass fuels indoors are even greater, with commensurate impacts on health.

A major regional impact of fossil fuel combustion is wet and dry deposition of sulfur and nitrogen, much of it in acidic forms. Of the sulfur oxide and nitrogen oxide emissions that are the precursors of this fallout, the former are somewhat easier to control technologically. Global emissions of both are now increasing, however, as rapid expansion of poorly controlled sources in Asia, and to a lesser extent in Africa and Latin America, is now more than offsetting reductions in the industrialized countries (11).

Mid-range projections for energy growth over the next few decades show world use of energy reaching 1.5 and 2 to 2.5 times the 2005 level by 2030 and 2050, respectively. Electricity generation in these "business-as-usual" cases nearly doubles by 2030 and triples by 2050 (13). Although these are daunting numbers from the standpoint of sustainability, the problem is not that the world is running out of energy. It isn't (2, 14). But it is running out of cheap and easy oil and gas, and it is running out of environmental capacity to absorb, without intolerable consequences, the impacts of mobilizing these quantities of energy in the ways we have been accustomed to doing it (15).

Much discussion of the oil issue has been framed around the contentious question of "peak oil" (16): when will global production of conventional petroleum reach a peak and begin to decline, as U.S. domestic production did around 1970? The question derives its importance from the proposition that reaching this peak globally will presage large and long-lasting increases in the price of oil, plus a costly and demanding scramble for alternatives to fill the widening gap between the demand for liquid fuel and the supply of conventional petroleum.

Oil-supply pessimists argue that the peak of conventional oil production could occur any time now; oil-supply optimists say it probably won't happen until after 2030, perhaps not until after 2050. Similar arguments go on about conventional supplies of natural gas, the total recoverable resources of which are thought to be not greatly different, in terms of energy content, from those of crude petroleum.

In my judgment, it's difficult to tell at this juncture whether the optimists or the pessimists are closer to right about when the world will experience peak oil, but the answer is not very important as a determinant of what we need to be doing. After all, it's clear that heavy dependence on oil carries substantial economic and political risks in a world where high proportions of the reserves and remaining recoverable resources lie in regions that are unstable and/or controlled by authoritarian governments that have sometimes been inclined to wield oil supply as a weapon. It's also clear that world oil use (which is dominated by the transport sector and, within it, by motor vehicles) is a huge producer of conventional air pollutants, as well as being about equal to coal burning as a contributor to the global buildup of the heat-trapping gas CO_2 (8, 11). Given these liabilities, it makes sense to be looking urgently for ways to reduce oil dependence (while working to clean up continuing uses of oil), no matter when we think peak oil might occur under business as usual.

Indeed, the problem of how to reduce the dangers from urban and regional air pollution and from overdependence on oil in the face of rising worldwide demand for personal transportation is one of the two greatest challenges at the energy-economy-environment intersection. The other one is how to provide the affordable energy needed to create and sustain prosperity everywhere without wrecking the global climate with the CO_2 emitted by fossil fuel burning.

Climate is the envelope within which nearly all other environmental conditions and processes important to human well-being must function (17). Climate strongly influences (so climate change directly affects) the availability of water; the productivity of farms, forests, and fisheries; the prevalence of oppressive heat and humidity; the geography of disease; the damages to be expected from storms, floods, droughts, and wildfires; the property losses to be expected from sea level rise; the investments of capital, technology, and energy devoted to ameliorating aspects of climate we don't like; and the distribution and abundance of species of all kinds (those we love and those we hate). A sufficient distortion in the climatic envelope, as recent human activities are well on the way to achieving, can be expected to have substantial impacts in most of these dimensions.

Indeed, after a rise in global average surface temperature of about 0.75° ± 0.20°C since 1880–1900 (18, 19), changes in most of these categories, and significant damages in many, have already become apparent (20, 21). Large impacts from seemingly modest changes in global average surface temperature underline

the reality that this temperature is a sensitive proxy for the state of the world's climate, which consists of the patterns in space and time not only of temperature and humidity but also of sun and clouds, rainfall and snowfall, winds and storm tracks, and more. (The sensitivity of the temperature proxy for the state of the climate is often illustrated by the observation that the difference in global average surface temperature between an ice age and a warm interglacial—drastically different climates—is only about 5°C.)

There is no longer any serious doubt that most of the climatic change that has been observed over the past few decades has been due to human rather than natural influences (22). As shown in Table 2, the largest of the positive human "forcings" (warming influences) has been the buildup of CO_2 in the atmosphere over the past two and a half centuries. (About two-thirds of this buildup has come from fossil fuel burning, and the other one-third, from land-use change.) Other important contributors have been methane from energy supply, land-use change, and waste disposal; halocarbons from a variety of commercial and industrial applications; nitrous oxide from fertilizer and combustion; and soot from inefficient engines and biomass burning. Partially offsetting cooling effects have been caused by the reflecting and cloud-forming effects of human-produced particulate matter and by increased surface reflectivity due to deforestation and desertification.

Facing the menace of growing, human-caused disruption of global climate, civilization has only three options: mitigation (taking steps to reduce the pace and the magnitude of the climatic changes we are causing), adaptation (taking steps to reduce the adverse impacts of the changes that occur), and suffering from impacts not averted by either mitigation or adaptation. We are already doing some of each and will do more of all, but what the mix will be depends on choices that society will make going forward. Avoiding increases in suffering that could become catastrophic will require large

TABLE 2. Disrupting Earth's Climate

Cause of forcing	Magnitude of forcing (W/m^2)
Change in atmospheric concentration of	
Carbon dioxide	+1.66 (±0.17)
Methane	+0.55 (±0.07)
Halocarbons	+0.34 (±0.03)
Nitrous oxide	+0.16 (±0.02)
Tropospheric ozone	+0.35 (−0.10,+0.30)
Stratospheric ozone	−0.05 (±0.10)
Soot	+0.3 (±0.2)
Reflecting particles	−0.8 (±0.4)
Cloud-forming effect of particles	−0.7 (−1.1,+0.4)
Change in reflectivity of surface (albedo) due to	
Land-use change	−0.2 (±0.2)
Soot on snow	+0.1 (±0.1)
Change in solar irradiance	+0.12 (−0.06,+0.18)

[a]Exajoules. [b]Terawatt-hours.

IPCC estimates of principal human-produced and natural forcings since 1750. Forcings are essentially changes in Earth's energy balance, measured in watts per square meter of the planetary surface, with positive values denoting warming influences and negative values denoting cooling. The uncertainty range is given in parentheses. Large volcanic eruptions produce negative forcings of a few years' duration due to the particles they inject into the atmosphere, but they are not included in the table because no trend is evident in the size of this effect over time. Effects of the 11-year sunspot cycle are likewise not shown because they average out over time periods longer than that. Note that the IPCC's best estimate of the contribution of the net change in input from the Sun since 1750 is some 14 times smaller than that of the CO_2 (18).

increases in the efforts devoted to both mitigation and adaptation.

A 2007 report for the United Nations Commission on Sustainable Development focused on what to do and, emphasizing mitigation and adaptation equally, concluded that the chances of a "tipping point" into unmanageable degrees of climatic change increase steeply once the global average surface temperature exceeds 2°C to 2.5°C above the preindustrial level and that mitigation strategies should therefore be designed to avoid increases larger than that (20). Having a better-than-even chance of doing this means stabilizing atmospheric concentrations of greenhouse gases and particles at the equivalent of no more than 450 to 500 parts per million by volume (ppmv) of CO_2 (23, 24).

A mitigation strategy sufficient to achieve such stabilization will need to address methane, halocarbons, nitrous oxide, and soot as well as CO_2, but the largest and most difficult reductions from business-as-usual trajectories of future emissions are those needed for CO_2 itself. The difficulty in the case of CO_2 emissions from the energy system resides in the current 80% dependence of world energy supply on fossil fuels, the technical difficulty of avoiding release to the atmosphere of the immense quantities of CO_2 involved, and the long turnover time of the energy system capital stock (meaning that the shares of the different energy sources are hard to change quickly) (25). In the case of the 15% to 25% of global CO_2 emissions still coming from deforestation (essentially all of it now in the tropics), the difficulty is that the causes of this deforestation are deeply embedded in the economics of food, timber, biofuel, trade, and development and in the lack of valuation and marketization of the services of intact forests (26).

Stabilizing atmospheric CO_2 at 500 ppmv would be possible if global emissions from fossil fuel combustion in 2050 could be cut in half from the mid-range business-as-usual figure of 14 billion metric tons of carbon in CO_2 per year. Numerous studies of how reductions of this general magnitude might be achieved have been undertaken (27), and notwithstanding differences in emphasis, virtually all have shown that (i) such reductions are possible but very demanding to achieve; (ii) there is no single silver-bullet approach that can do all or even most of the job; (iii) it is essential, in terms of both feasibility of the ultimate aim and cost of achieving it, to begin reductions sooner rather than later; (iv) the quickest and cheapest available reductions will be through improving the efficiency of energy end-use in residential and commercial buildings, manufacturing, and transport, but costlier measures to reduce emissions from the energy supply system will also need to be embraced; and (v) without major improvements in technology on both the demand side and the supply side—and a major expansion of international cooperation in the development and deployment of these technologies—the world is unlikely to achieve reductions as large as required.

The improved technologies we should be pursuing, for help not only with the energy-climate challenge but also with other aspects of the energy-economy-environment dilemma, are of many kinds: improved batteries for plug-in hybrid vehicles; cheaper photovoltaic cells; improved coal gasification technologies to make electricity and hydrogen while capturing CO_2; new processes for producing hydrogen from water using solar energy; better means of hydrogen storage; cheaper, more durable, more efficient fuel cells; biofuel options that do not compete with food production or drive deforestation; advanced fission reactors with proliferation-resistant fuel cycles and increased robustness against malfunction and malfeasance; fusion; more attractive and efficient public transportation options; and a range of potential advances in materials science, biotechnology, nanotechnology, information technology, and process engineering that could drastically reduce the energy and resource requirements of manufacturing and food production (28).

Also urgently needed from science and technology in the energy-climate domain are improved understanding of potential tipping points related to ice sheet disintegration and carbon release from the heating of northern soils; a greatly expanded research, development, and demonstration effort to determine the best approaches for both geologic and enhanced biologic sequestration of CO_2; a serious program of research to determine whether there are "geoengineering" options (to create global cooling effects that counter the ongoing warming) that make practical sense; and wide-ranging integrated assessments of the options for adaptation (29).

Adequately addressing these and other needs in the science and engineering of the energy-environment interaction would probably require something like a 2- to 3-fold increase in funding for fundamental environmental science and a 4- to 10-fold increase in the sum of public and private spending for energy research, development, and demonstration (ERD&D) (30). Achieving such increases seems daunting in an era of budget constraints, but the amounts involved are small compared either with the potential environmental damages from climate change and other environmental impacts of energy or with what society spends for energy itself (31). There are signs that the private sector is ramping up its efforts in ERD&D in response to the challenge, but for reasons that have been abundantly documented (32), the public sector must also play a large role in the needed expansion. Sadly, until now there has been precious little sign of that happening, notwithstanding abundant rhetoric from political leaders about new technologies' being the key to the solution (33).

Despite the current inadequacies in funding, researchers at the frontiers of these problems continue to make major advances both in clarifying the character of the threats and in exploring the potential remedies. In the pages that follow, leading scientists tell you more about such progress, not only in relation to the energy-environment dilemma, but on other aspects of the intersection of science and technology with sustainable well-being, as well.

References and Notes

1. The current essay was excerpted and adapted from the presidential address, which was published

as John P. Holdren, Science and technology for sustainable well-being, *Science* 319, 5862 (2008).

2. M. K. Hubbert, in *Resources and Man,* National Research Council (W. H. Freeman, San Francisco, 1969), chap. 8.

3. J. Holdren, P. Herrera, *Energy* (Sierra Club Books, New York, 1971).

4. J. Goldemberg, Ed., *The World Energy Assessment* (UNDP, UN Department of Economic and Social Affairs, and World Energy Council, New York, 2000).

5. Much of this was already clear from the pioneering report of the 1970 summer workshop organized at the Massachusetts Institute of Technology (MIT) by Carroll Wilson, *Study of Critical Environment Problems* (MIT Press, Cambridge, MA, 1970). A more recent synoptic account is the chapter on "Energy, Environment, and Health," in J. P. Holdren, K. R. Smith, convening lead authors, in (4).

6. Data for Fig. 1 were compiled and reconciled from J. Darmstadter, *Energy in the World Economy* (Johns Hopkins Univ. Press, Baltimore, 1968); D. O. Hall, G. W. Barnard, P. A. Moss, *Biomass for Energy in Developing Countries* (Pergamon, Oxford, 1982); BP Amoco, *Stat. Rev. World Energy* (BP, London, annual); and (2). Graphic courtesy of S. Fetter.

7. J. P. Holdren, *Popul. Environ.* 12, 231 (1991).

8. International Energy Agency, *Key World Energy Statistics 2007* (OECD, Paris, 2007).

9. P. J. Crutzen, W. Steffen, *Clim. Change* 61, 251 (2003).

10. For earlier discussions of this issue, see, e.g., J. Holdren, P. Ehrlich, *Am. Sci.* 62, 282 (1974).

11. UNEP, *Global Environmental Outlook 4* (GEO-4, UNEP, Nairobi, Kenya, 2007).

12. C. A. Pope *et al.*, *JAMA* 287, 1132 (2002); J. Kaiser, *Science* 307, 1858 (2005).

13. U.S. Energy Information Administration, *International Energy Outlook 2007* (U.S. Department of Energy, Washington, DC, 2007).

14. See, e.g., IPCC, *Climate Change 2007: Mitigation. Working Group III Contribution to the IPCC Fourth Assessment Report* (IPCC, Geneva, 2007).

15. J. P. Holdren, *Innovations* 1, 3 (2006).

16. Credit for the idea of approximating the production trajectory of depletable resources as a Gaussian curve and for insights about the significance of the peak year and how to predict it belongs to the late geophysicist M. King Hubbert, who in the 1950s used this approach to correctly predict that U.S. domestic production of conventional oil would peak around 1970 [(2) and references therein]. He also predicted that world production of crude petroleum would peak between 2000 and 2010. Reviews, extensions, and critiques of Hubbert's approach now constitute a considerable literature; see, e.g., K. Deffeyes, *Hubbert's Peak: The Impending World Oil Shortage* (Farrar, Straus & Giroux, New York, 2002) and C. J. van der Veen, *Eos* 87, 199 (2006).

17. Some of the formulations about climate in what follows have been adapted from (15).

18. IPCC, *Climate Change 2007: The Physical Science Basis, Contribution of Working Group I to the Fourth Assessment Report of the IPCC* (Cambridge Univ. Press, Cambridge, 2007).

19. The beginning of the buildup of atmospheric greenhouse gases attributable to human activities dates back to even before 1750, the nominal start of the Industrial Revolution and the zero point used by the IPCC for its estimates of subsequent human influences. Earlier human contributions to atmospheric greenhouse gas concentrations came principally from deforestation and other land-use change (9). The human influences on global average surface temperature did not become large enough to be clearly discernible against the backdrop of natural variability until the 20th century, however. See especially J. Hansen *et al.*, *Proc. Natl. Acad. Sci. USA* 103, 14288 (2006), as well as (18).

20. P. Raven *et al.*, *Confronting Climate Change: Avoiding the Unmanageable and Managing the Unavoidable* (UN Foundation, Washington, DC, 2007).

21. IPCC, *Climate Change 2007: Impacts, Adaptation, and Vulnerability, Contribution of Working Group II to the Fourth Assessment Report of the IPCC* (Cambridge Univ. Press, Cambridge, 2007); UNDP, *Human Development Report 2007–2008: Fighting Climate Change* (UNDP, Washington, DC, 2007).

22. Even the IPCC, which by its structure and process is designed to be ultraconservative in its pronouncements, rates the probability that most of the observed change has been due to human influences as between 90% and 95% in its 2007 report (18).

23. For convenience, the IPCC and other analysts often represent the net effect of all of the human influences on Earth's energy balance as the increased concentration of CO_2 alone that would be needed to achieve the same effect, starting from a reference point of 278 ppmv of CO_2 in 1750. In 2005, when the actual CO_2 concentration was 379 ppmv, the additional warming influences of the non-CO_2 greenhouse gases and soot were the equivalent of another 100 ppmv of CO_2, and the cooling effects of human-produced reflecting and cloud-forming particles and surface reflectivity changes were (coincidentally) equivalent to subtracting about the same amount of CO_2. Thus, the net effect was about what would have been produced by the actual CO_2 increase alone (see Table 2).

24. The relationship between climate forcing (represented as the CO_2 concentration increase that would give the same effect as all of the human influences combined) and the corresponding change in global average surface temperature must be expressed

in probabilistic terms because of uncertainty about the value of climate "sensitivity," which is commonly defined as the temperature change that would result from forcing that corresponds to a doubling of the 1750 CO_2 concentration. See especially S. Schneider, M. Mastrandrea, *Proc. Natl. Acad. Sci. USA* 102, 15728 (2005) as well as (*18*).

25. About 27.5 billion tons of CO_2, containing 7.5 billion tons of carbon, were emitted by fossil fuel combustion in 2005. The replacement cost of the current world energy system is in the range of $15 trillion, and the associated capital stock has an average turnover time of at least 30 to 40 years. See, e.g., International Energy Agency, *World Energy Outlook 2006* (OECD, Paris, 2006) and (*20*).

26. P. Moutinho, S. Schwartzman, Eds., *Tropical Deforestation and Climate Change* (Instituto de Pesquisa Ambiental da Amazônia, Belem, and Environmental Defense, Washington, DC, 2005).

27. M. Hoffert *et al.*, *Science* 298, 981 (2002); S. Pacala, R. Socolow, *Science* 305, 968 (2004); P. Enkvist, T. Nauclér, J. Rosander, *McKinsey Quart.* 1, 35 (2007); J. Edmonds *et al.*, *Global Energy Technology Strategy* (Battelle Memorial Institute, Washington, DC, 2007) and (*47*).

28. See., e.g., N. Lane, K. Matthews, A. Jaffe, R. Bierbaum, Eds., *Bridging the Gap Between Science and Society* (James A. Baker III Institute for Public Policy, Rice Univ., Houston, TX, 2006).

29. D. W. Keith, *Annu. Rev. Energy Environ.* 25, 245 (2000); P. J. Crutzen, *Clim. Change* 77, 211 (2006); and (*20*)

30. See, e.g., President's Committee of Advisors on Science and Technology, *Federal Energy Research and Development for the Challenges of the 21st Century* (Executive Office of the President of the United States, Washington, DC, 2007); World Energy Council (WEC), *Energy Technologies for the 21st Century* (WEC, London, 2001); National Commission on Energy Policy (NCEP), *Breaking the Energy Stalemate* (NCEP, Washington, DC, 2004); and G. F. Nemet, D. M. Kammen, *Energy Policy* 35, 746 (2007).

31. Expenditures of firms and individuals for energy are generally in the range of 5% to 10% of gross domestic product—in round numbers, perhaps a trillion dollars per year currently in the United States and five times that globally. Estimates of expenditures by governments on ERD&D depend on assumptions about exactly what should be included but, by any reasonable definition, are currently not more than $12 billion to $15 billion per year worldwide. Private-sector investments in ERD&D are much more difficult to estimate. If following the general pattern in the United States, they are assumed to be twice government investments; then the public-plus-private total for the world is in the range of $35 billion to $50 billion per year, which is equal to at most 1% of what is spent on energy itself. By contrast, many other high-technology sectors spend 8% to 15% of revenues on R&D [see (*30*)].

32. See, e.g., K. S. Gallagher, J. P. Holdren, A. D. Sagar, *Annu. Rev. Environ. Resources* 31, 193 (2006); President's Committee of Advisors on Science and Technology, *Powerful Partnerships: The Federal Role in International Cooperation on Energy-Technology Innovation* (Executive Office of the President of the United States, Washington, DC, 1999); and (*30*).

33. K. S. Gallagher, A. D. Sagar, D. Segal, P. de Sa, J. P. Holdren, *DOE Budget Authority for Energy Research, Development, and Demonstration Database* (Energy Technology Innovation Project, Cambridge, MA, 2006).

Renewable Energy Sources and the Realities of Setting an Energy Agenda

JANEZ POTOČNIK

Energy is undoubtedly moving up the political agenda as an issue that needs to be addressed urgently. If last year's threats to European gas supplies during the dispute between Russia and Ukraine did not show the immediacy of the challenges such as energy supply, then the report toward the end of last year by Sir Nicholas Stern (1) on the economics of climate change must surely have rung a warning bell.

The European Commission has been devoting considerable attention to energy issues for some time now. We were leaders in the process that brought about the Kyoto Protocol and have developed the first large-scale emissions trading scheme in the world. In March 2006, we published a Green Paper on energy (2), which

This article first appeared in *Science* (9 February 2007: Vol. 315, no. 5813). It has been revised for this edition.

we have now, at the beginning of 2007, followed up with a strategic energy package (3) addressing energy policy in general and also outlining future European policy on various specific elements.

One of these specific elements will be the elaboration at the European level of a Strategic Energy Technology Plan (4). Research and technology will undoubtedly be crucial to cracking the energy and climate change nut. A recent study published by the European Commission (see image) (5) shows that if existing trends continue, by 2050, CO_2 emissions will be unsustainably high: 900 to 1000 **parts per million** by volume, that is, well above what is considered an acceptable range for stabilization. Without determined action, energy demand will double and electricity demand will quadruple, resulting in an 80% increase in CO_2 emissions. However, technological development coupled with strong carbon constraint policies can limit this impact, with world emissions stable between 2015 and 2030 and decreasing thereafter. In this "carbon constraint" case, half the total building stock would be made of low-energy buildings, and more than half of the vehicles would have low or very low emissions, a clear example of how technological development will contribute to our energy and environmental policy objectives.

The strategic energy package sets a target of 20% of Europe's energy coming from renewable sources by 2020. If successful, this would mean that by 2020, the European Union (EU) would use about 13% less energy than today, saving €100 billion and around 780 metric tons of CO_2 each year. For this to be realistic, significant strides need to be made, technologically speaking. Today renewable energy is, on the whole, costly and intermittent. Even if we are looking to maintain a mix of sources of energy, a cloudy, windless day rules out generation from solar and wind power. And yet, on a bright, windy day, energy may go unused because it cannot be stored easily. Reliability and continuity are basic requirements if renewable sources of energy are to be seen as viable alternatives to oil, gas, and coal. Research and technological development are already bringing us closer to solutions in this field, through improving fuel cells or redesigning electricity grids to deal with more decentralized power generation.

We believe that renewables have the potential to provide around a third of EU electricity by 2020 (3). Current statistics indicate that this is not an unreasonable goal. Wind power currently provides roughly 20% of electricity needs in Denmark, as well as 8% in Spain and 6% in Germany. If other Member States matched

the levels that Sweden, Germany, and Austria have attained in geothermal heat pumps and solar heating, the share of renewable energy in heating and cooling would jump by 50%. As for biofuels, Sweden has already achieved a market share of 4% of the petrol market for bioethanol, and Germany is the world leader for biodiesel, with 6% of the diesel market. Biofuels could account for as much as 14% of transport fuels by 2020 (3). The European public is also clearly in favor of advancing renewable sources of energy, with a recent opinion poll (6) showing approval ratings for such energy ranging between 55% and 80%.

The European Commission has certainly taken this on board in its new research funding program, the Seventh Framework Programme (7). Within the energy theme of the cooperation program, which will focus on noncarbon or reduced-carbon sources of energy, emphasis will be given to renewable electricity generation, renewable fuel production, hydrogen and fuel cells, CO_2 capture and storage technologies, **smart energy networks**, energy efficiency and savings, nuclear fission safety and waste management, the development of **fusion energy**, and knowledge for energy policymaking. The Seventh Framework Programme increases the annual funding available to energy research at the European level to €886 million a year, compared with €574 million a year in the previous program. But this is not enough; more combined effort is needed. In some areas, we have moved toward common research agendas at the

Science, Vol. 315, no. 5813, 790, 9 February 2007

STEVEN KOONIN PROFILE:
GUIDING AN OIL TANKER INTO RENEWABLE WATERS
Daniel Clery

LONDON—Steven Koonin's career took a surprising turn in 2004. After nearly three decades as an academic theoretical physicist, including nine years as provost of the California Institute of Technology (Caltech) in Pasadena, he pulled up his Southern California roots and moved to London to become chief scientist of the oil company BP. But his new role is not all about oil. BP was the first major oil company to acknowledge publicly that people may be causing climate change. Now, committed to moving "beyond petroleum," it is complementing oil exploration with investment in solar and wind power, clean coal technology, and biofuels. Koonin's job is to help allocate BP's $500-million-a-year research funding, plot its technology strategy, and generally evangelize about the energy challenges facing the world. "Sometimes I feel like I've got the most wonderful job in the world," Koonin says.

Alive to the challenge.
Koonin wants to "do biofuels right."
CREDIT: S. KOONIN

Koonin says he spent his first year and a half in the job learning about energy, a process that changed his views. "I was more skeptical about climate change a few years ago. Now I've come round more toward the IPCC view." (The Intergovernmental Panel on Climate Change has concluded that temperatures are rising in part as a result of human activity.) Like most energy pundits, he sees no silver bullet that will save the planet from climate change, but "some ammo has a bigger caliber than others." He advocates large-scale efforts in carbon sequestration at fossil fuel–burning plants, as well as a new generation of nuclear power stations. And he says it would be "irresponsible" not to investigate other ways to deal with global warming, such as geoengineering: increasing Earth's reflectivity by pumping material into the upper atmosphere.

One area claiming much of his attention is biofuels. Current biofuel efforts, he argues, are strapped onto agricultural food production and are "not optimal." Koonin wants to galvanize geneticists, biotechnologists, agricultural scientists, engineers, and others to "do biofuels right." BP has put up $500 million over 10 years to create an Energy Biosciences Institute (EBI).

This initiative has won plaudits. "Koonin thought hard about how to structure the EBI, and it will have a lot of impact," says carbon sequestration expert Robert Socolow of Princeton University. "BP has made a commitment to go big in energy biosciences. I doubt this would have happened without Steve Koonin," says Ernest Moniz, Professor of Physics at MIT.

Koonin is pleased with the buzz the EBI has caused. "Plant geneticists are talking to chemists and engineers.... Researchers are coming alive to the challenge," he says.

European level through the creation of European technology platforms (8). Several exist in the energy field, including for hydrogen and fuel cells, photovoltaics, zero-emission fossil fuel power plants, and **smart grids**. Nonetheless, we have seen investment in energy research being reduced in national budgets over the past 20 years or so. And the research that is carried out is more often than not done in a fragmented, uncoordinated way, leading to duplication in some areas and to other important aspects being underfunded or ignored. This is the raison d'être of the Strategic Energy Technology Plan, which will, once agreed on, provide a basis for all energy technology efforts in Europe, overcoming the lack of coherence that has unfortunately been present to a greater or lesser extent in the research programs at the national and European levels up to now.

During the first half of 2007, the European Commission will consult intensively with all those who have a role to play in such a strategic plan. On the basis of these consultations, a text will be drawn up toward the middle of the year, upon which the research community, among others, will be invited to give its comments. It is important that the creation of the Strategic Energy Technology Plan be a collaborative, bottom-up process if it is to have any chance of

KEY TERM

Fusion energy is generated by nuclear fusion reactions, in which two light atomic nuclei (isotopes of hydrogen) fuse together to form a heavier nucleus (helium) and, in doing so, release large amounts of energy. Fusion is the fundamental energy source of the universe and is the process that powers the Sun and the stars.

KEY TERM

Smart grids and smart energy networks are electrical transmission networks that use advanced sensing, communication, and control technologies to distribute electricity more efficiently and economically than traditional electrical grids.

achieving its stated objective of being a reference point for future EU activities in this area.

Since my appointment as European Science and Research Commissioner in November 2004, I have insisted on the importance of science and research as the key to solving many of the challenges that we face. I can think of no better illustration of this approach than the issue of energy. Here, we have various requirements in front of us: finding secure and sustainable sources of energy that support our economic growth and competitiveness without damaging our environment. The answer to reconciling these requirements lies in knowing more and being better. We have a chance to work together to develop solutions to the problems of climate change and energy supply that not only ensure our future economic development, but give European scientists and companies the opportunity to be (or remain) at the cutting edge of technological development. It is crucially important that we take this opportunity and make it work (9).

References and Notes

1. *Stern Review on the Economics of Climate Change* (2006); www.hm-treasury.gov.uk/independent_reviews/stern_review_economics_climate_change/sternreview_index.cfm.

2. Commission of the European Communities, *Green Paper: A European Strategy for Sustainable, Competitive and Secure Energy*, SEC (2006) 317;

http://ec.europa.eu/energy/green-paper-energy/doc/
.2006_03_08_gp_document_en.pdf.

3. Commission of the European Communities,
*Communication from the Commission to the European
Council and the European Parliament: An Energy Policy
for Europe*, COM (2007) 1 final; http://ec.europa.eu/
energy/energy_policy/doc/01_energy_policy_ for_
europe_en.pdf.

4. European Commission, *Toward A Strategic
Energy Technology Plan*, COM (2007) 847; ec.europa
.eu/energy/energy-policy/doc/19_strategic_energy_
technology_plan_en.pdf.

5. European Commission, *World Energy Technology

Outlook—2050* (2006); http://ec.europa.eu/research/
energy/pdf/weto-h2_en.pdf.

6. European Commission, *Energy Technologies:
Knowledge-Perception-Measures* (2006); http://ec
.europa.eu/research/energy/pdf/energy_tech_euro
barometer_en.pdf.

7. European Commission, *Official Journal*,
document L412 (2006); http://eur-lex.europa.eu/
LexUriServ.do?uri=OJ:L:2006:412:0001:EN:HTML.

8. For further information, see http://cordis
.europa.eu/technology-platforms/individual_en.html.

9. The author is the European Commissioner for
Science and Research.

Ethanol for a Sustainable Energy Future

JOSÉ GOLDEMBERG

R enewable energy is one of the most efficient ways to achieve sustainable development. Increasing its share in the world matrix will help prolong the existence of fossil fuel reserves, address the threats posed by climate change, and enable better security of the energy supply on a global scale. Most of the "new renewable energy sources" are still undergoing large-scale commercial development, but some technologies are already well established. These include Brazilian sugarcane ethanol, which, after 30 years of production, is a global energy commodity that is fully competitive with motor gasoline and appropriate for replication in many countries.

A sustainable energy future depends on an increased share of renewable energy, especially in developing countries. One of the best ways to achieve such a goal is by replicating the large Brazilian program of sugarcane ethanol, started in the 1970s.

The World Commission on Environment and Development (WCED) in 1987 defined "sustainable development" as development that "meets the needs of the present without compromising the ability of future generations to meet their own needs" (1). The elusiveness of such a definition has led to unending discussions among social scientists regarding the meaning of "future generations."

However, in the case of energy, exhaustible fossil fuels represent about 80% of the total world energy supply. At constant production and consumption, the presently known reserves of oil will last about 41 years, natural gas 64 years, and coal 155 years (2). Although very simplified, such an analysis illustrates why fossil fuels cannot be considered as the world's main source of energy for more than one or two generations.

This article first appeared in *Science* (9 February 2007: Vol. 315, no. 5813). It has been revised for this edition.

Besides the issue of depletion, use of fossil fuels presents serious environmental problems, particularly global warming. Also, their production costs will increase as reserves approach exhaustion and as more expensive technologies are used to explore and extract less attractive resources. Finally, there are increasing concerns for the security of the oil supply, originating mainly from politically unstable regions of the world.

Except for nuclear energy, the most likely alternatives to fossil fuels are renewable sources such as hydroelectric, biomass, wind, solar, geothermal, and marine tidal energy. Figure 1 shows the present world energy use.

Fossil fuels (oil, coal, and gas) represent

SCIENCE IN THE NEWS

*Science*NOW Daily News, 12 March 2008

NITRATE THREATENING THE NATION'S WATERSHEDS
Phil Berardelli

There's mixed news about how the country's streams and rivers are handling increased loads of nitrate from human activities. The ecosystems are normally highly tolerant of the chemical, which is good. But new research shows that nitrate absorption can reach a limit, and that's what is happening in many areas. Worse, the budding biofuels industry figures to release even more nitrate into watersheds in the coming years.

Livestock waste and nitrogen-based fertilizers constitute the biggest source of nitrate in U.S. watersheds. The chemical, also known as NO_3, works its way into streams and rivers, where it feeds nitrogen-loving plants and bacteria. Eventually, some of it ends up in bays and estuaries, nourishing plants and algae there as well. But in large amounts, nitrate can unbalance ecosystems, fueling rampant growth by plants and algae. Their subsequent decomposition by bacteria can pull so much oxygen out of the water that it causes a condition called hypoxia, which can stress fish and other marine creatures. In the worst cases, the oxygen can disappear altogether, resulting in dead zones and spawning dangerous algal blooms called red tides (*Science*NOW, 28 June 2006).

Scientists have long known that plants and bacteria in watersheds serve as a nitrate disposal system, but up to now, no one has been able to gauge how efficiently the system works. So a team of researchers from 15 states set out to study the process in action. They dumped nitrate spiked with a harmless nitrogen isotope as a marker into 72 streams across the continental United States and Puerto Rico. Then they collected samples at various points downstream continuously over a 24-hour period to see how well nitrate was absorbed.

"Our results confirmed our initial suspicions in that we found relatively high removal rates of nitrate from stream water in most streams," says aquatic ecologist Patrick Mulholland of Oak Ridge National Laboratory in Tennessee. Stream ecosystems were able to remove between 10% and 20% of the added nitrate within an hour. But the big surprise, Mulholland says, was that increasing nitrate concentrations caused a precipitous drop in absorption efficiency—meaning much more nitrate could make its way downstream if nitrate loads continue to increase.

The most likely cause of heavier nitrate loads in the future, Mulholland says, is expanded biofuels production, which could require large increases in nitrogen-based fertilizers. "These problems are likely to be most severe for biofuel crops such as corn," he says, but may be less severe for switchgrass and other perennials. "It's a difficult problem to quantify."

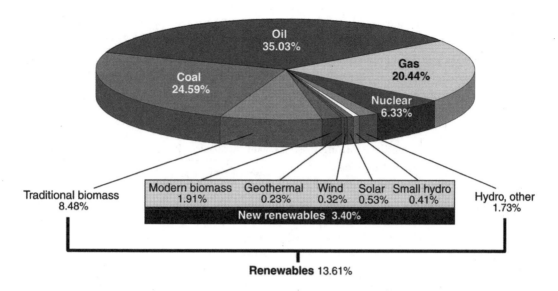

FIGURE 1: World total primary energy supply 2004,
shares of 11.2 billion tons of oil equivalent, or 470 EJ (15, 16).

80.1% of the total world energy supply, nuclear energy 6.3%, and renewables 13.6%. The largest part of renewables is **traditional biomass** (8.5% of total primary energy), which is used mainly in inefficient ways, such as in highly polluting primitive cooking stoves used by poor rural populations, leading in many cases to deforestation.

The "new renewable energy sources" amount to 16 exajoules (1 EJ = 10^{18} J), or 3.4% of the total. Table 1 shows a breakdown of the contribution of new renewables, which include small hydropower plants. Many of these technologies, including solar, wind, geothermal, and **modern biomass**, are still undergoing large-scale commercial development. The largest part (1.9% of the total) is modern biomass, which refers to biomass produced in a sustainable way and used for electricity generation, heat production, and transportation of liquid fuels. It includes wood and forest residues from reforestation and/or sustainable management, as well as rural (animal and agricultural) and urban residues (including solid waste and liquid effluents).

From the perspective of sustainable energy development, renewables are widely available, ensuring greater security of the energy supply and reducing dependence on oil imports from politically unstable regions. Renewables are less polluting, both in terms of local emissions (such

KEY TERMS

Traditional biomass is unprocessed biomass, including agricultural waste, forest products waste, collected fuel wood, and animal dung, that is burned in stoves or furnaces to provide heat energy for cooking, heating, and agricultural and industrial processing, typically in rural areas.

Modern biomass refers to technologies other than those defined for traditional biomass, such as biomass cogeneration for power and heat, biomass gasification, biomass anaerobic digesters, and liquid biofuels for vehicles.

as particulates, sulfur, and lead) and greenhouse gases (carbon dioxide and methane) that cause global warming. They are also more labor-intensive, requiring more workforce per unit of energy than conventional fossil fuels (3).

Although technologically mature, some of the renewable sources of energy are more expensive than energy produced from fossil fuels. This is particularly the case for the "new renewables." Traditional biomass is frequently not the object of commercial transaction, and it is difficult to evaluate its costs, except the environmental ones. Cost continues to be the fundamental barrier to widespread adoption of traditional biomass despite its attractiveness from a sustainability perspective.

A number of strategies have been adopted by governments in the industrialized countries and international financial institutions to encourage the use of "new renewables," and there have been several successes, based on the use of tax breaks, subsidies, and renewable portfolio standards (RPS). Examples are the large growth (of more than 35% per year, "albeit" from a low base value) for wind and solar photovoltaics in industrialized countries such as Denmark, Germany, Spain, and the United States (4). These technologies are slowly spreading to developing countries through several strategies.

In developing countries, the best example of large growth in the use of renewables is given by the sugarcane ethanol program in Brazil. Today, ethanol production from sugarcane in the country is 16 billion liters (4.2 billion gal) per year, requiring around 3 million hectares (ha) of land. The competition for land use between food and fuel has not been substantial: Sugarcane covers 10% of total cultivated land and 1% of total land available for agriculture in the country. Total sugarcane crop area (for sugar and ethanol) is 5.6 million ha.

Production of ethanol from sugarcane can be replicated in other countries without serious damage to natural ecosystems. Worldwide, some 20 million ha are used for growing sugarcane, mostly for sugar production (5). A simple

TABLE 1. "New renewables," by source in 2004 (15); updated with data from (4, 16). Assumed average conversion efficiency: for biomass heat, 85%; biomass electricity, 22%; biomass combined heat and power (CHP), 80%; geothermal electricity, 10%; all others, 100%.

Source/technology	2004	
	Exajoules (EJ)	Share in this sector
Modern biomass energy		
Total	9.01	56.19%
Bioethanol	0.67	
Biodiesel	0.07	
Electricity	1.33	
Heat	6.94	
Geothermal energy		
Total	1.08	6.77%
Electricity	0.28	
Heat	0.80	
Small hydropower		
Total	1.92	12.00%
Wind electricity		
Total	1.50	9.35%
Solar		
Total	2.50	15.63%
Hot water	2.37	
Photovoltaic electricity, grid	0.06	
Photovoltaic electricity, off-grid	0.06	
Thermal electricity	0.01	
Marine energy (tidal)		
Total	0.01	
Total	16.03	100.00%

calculation shows that expanding the Brazilian ethanol program by a factor of 10 (i.e., an additional 30 million ha of sugarcane in Brazil and in other countries) would supply enough ethanol to replace 10% of the gasoline used in the world. This land area is a small fraction of the more than 1 billion ha of primary crops already harvested on the planet.

What was the process that established firmly the ethanol program in Brazil? In the late 1970s, the Brazilian Federal Government mandated the mixture of anhydrous ethanol in gasoline (blends up to 25%) and encouraged car makers to produce engines running on pure hydrated ethanol (100%). Brazilian adoption of manda-

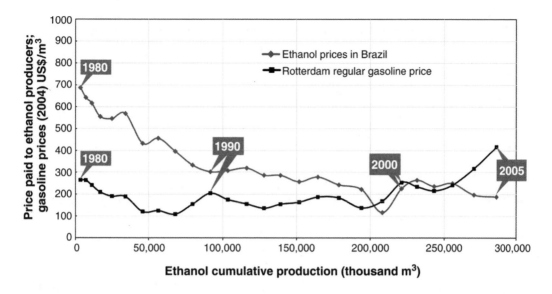

FIGURE 2. Ethanol learning curve in volume, comparing the price paid to ethanol producers in Brazil with the price of gasoline in the international market of Rotterdam (6).

tory regulations determining the amount of ethanol to be mixed with gasoline (basically a renewable portfolio standard for fuel) was essential to the success of the program. The motivation was to reduce oil imports, which were consuming one-half of the total amount of hard currency from exports. Although it was a decision made by the federal government during a military regime, it was well accepted by the civil society, agricultural sector, and car manufacturers. Similar policies are being considered by the European Union, Japan, and several states in the United States.

Such a policy decision created a market for ethanol, and production increased rapidly. Ethanol costs declined along a "learning curve" (6) as production increased an average 6% per year, from 0.9 billion gal in 1980 to 3.0 billion gal in 1990 and to 4.2 billion gal in 2006. The cost of ethanol in 1980 was approximately three times the cost of gasoline, but governmental cross-subsidies paid for the price difference at the pump. The subsidies came mostly from taxes on gasoline and were thus paid by automobile drivers. All fuel prices were controlled by the government. Overall subsidies to ethanol are estimated to be around US$30 billion over

20 years (7), but they were more than offset by a US$50 billion reduction of petroleum imports as of the end of 2006. Since the 1990s, subsidies have been progressively removed, and by 2004, ethanol became fully competitive with gasoline on the international markets without government intervention. Subsidies for ethanol production are a thing of the past in Brazil (Fig. 2) because new ethanol plants benefit from the economies of scale and the modern technology available today, such as the use of high-pressure boilers that allow cogeneration of electricity, with surpluses sold to the electric power grid.

The Brazilian ethanol program started as a way to reduce the reliance on oil imports, but it was soon realized that it had important environmental and social benefits (8). Conversion to ethanol allowed the phaseout of lead additives and MTBE (methyl tertiary butyl ether), and it reduced sulfur, particulate matter, and carbon monoxide emissions. It helped mitigate greenhouse gas emissions efficiently, by having a net positive energy balance; also, sugarcane ethanol in Brazil costs less than other present technologies for ethanol production (Table 2) and is competitive with gasoline in the United States, even considering the import duty of US$0.54

per gallon and energy efficiency penalties (30% or less with modern flexible fuel vehicle technologies) (9). The summer wholesale price of gasoline in the United States is about $1.9 per gallon; the corn ethanol wholesale price is about US$2.5 per gallon (10). Cellulose ethanol is a promising option in the long term, but it is not being produced on a commercial scale. The longer-term target is as low as 60 cents per gallon, but this will require major advances in producing, collecting, and converting biomass. A more realistic research target is to reduce the cost of production to US$1.07 per gallon until 2012 (11).

The development of other biomass-derived fuels in Brazil or elsewhere could benefit from such insights. Promising candidates along those lines are the following:

- Ethanol can be he produced from cellulosic materials, though considerable R&D (research and development) effort is still required before it reaches the production stage. If the technology for such conversion is firmly established, it would open enormous opportunities for the use of all kinds of wood and other biomass feedstocks for ethanol production.

- The enhanced use of biogas produced from microbial conversion in landfills of municipal solid wastes, wastewater, industrial effluents, and manure wastes will abate a considerable share of greenhouse gases that would be released to the atmosphere, replacing also fossil fuels for heat and electricity production.

KEY TERM

A fuel's energy balance is the ratio of the energy contained in a given volume of the fuel to the fossil energy expended to produce it. It is an indicator of the renewable character of the fuel.

TABLE 2. Ethanol costs and energy balances.

Feedstock	Cost (US$ per gallon)	Energy balance (renewable output to fossil input)
Sugarcane, Brazil		10.2 (18)
2006, without import tax	0.81 (17)	
2006, with U.S. import tax	1.35 (9, 17)	
Sugar beet, Europe, 2003	2.89 (17)	2.1 (19)
Corn, U.S., 2006	1.03 (17)	1.4 (9, 11)
Cellulose ethanol, U.S.		10.0 (11)
Achieved in 2006	2.25 (11)	
Target for 2012	1.07 (11)	

- The use of planted forests for the production of electricity either by direct combustion or by gasification and use of highly efficient gas turbines will also replace efficiently coal, natural gas, oil, and even nuclear sources. Reforested wood can also reduce the need for deforested fuelwood, controlling efficiently releases of greenhouse gases through market-friendly initiatives.

The ethanol program in Brazil was based on indigenous technology (in both the industrial and agricultural areas) and, in contrast to wind and solar photovoltaics, does not depend on imports, and the technology can be transferred to other developing countries.

Until breakthrough technologies become commercially viable, an alternative already exists: Many developing countries have suitable conditions to expand and replicate the Brazilian sugarcane program, supplying the world's gasoline motor vehicles with a renewable, efficient fuel (20).

References and Notes

1. United Nations, *Report of the World Commission on Environment and Development*, United Nations General Assembly, 96th plenary meeting, 11 December 1987, Document A/RES/42/187; www.un.org/documents/ga/res/42/ares42-187.htm.

2. British Petroleum, *BP Statistical Review of World Energy;* www.bp.com/liveassets/bp_internet/ globalbp/globalbp_uk_english/publications/energy_ reviews_2006/STAGING/local_assets/downloads/ spreadsheets/statistical_review_full_report_workbook _2006.xls.

3. J. Goldemberg, *The Case for Renewable Energies* (background paper for the International Conference for Renewable Energies, Bonn 2004); www.renewables 2004.de/pdf/tbp/TBP01-rationale.pdf. On jobs, see also (12).

4. REN21, *Global Status Report 2006 Update* (Renewable Energy Policy Network for the 21st Century, 2006); www.ren21.net/pdf/RE_GSR_ 2006_Update.pdf.

5. FAO, FAOSTAT (United Nations Food and Agriculture Organization, 2006); http://faostat.fao .org/default.aspx.

6. J. Goldemberg, S. T. Coelho, O. Lucon, P. M. Nastari, *Biomass Bioenergy* 26, 301 (2004).

7. J. Goldemberg, S. T. Coelho, O. Lucon, *Energy Policy* 32, 1141 (2004).

8. J. G. Da Silva, G. E. Serra, J. R. Moreira, J. C. Conçalves, J. Goldemberg, *Science* 201, 903 (1978).

9. S. T. Coelho, J. Goldemberg, O. Lucon, P. Guardabassi, *Development* 10, 26 (2006). On ethanol duties, see also (13).

10. J. R. Healey, "Ethanol shortage could up gas prices," *USA Today*, 30 March 2006; www.usatoday .com/money/industries/energy/2006-03-30-ethanol- gas-prices_x.htm.

11. M. Pacheco, *U.S. Senate Full Commitee Hearing—Renewable Fuel Standards* (National Renewable Energy Laboratory, National Bioenergy Center, 19 June 2006); http://energy.senate.gov/ public/index.cfm?IsPrint=true&FuseAction= Hearings.Testimony&Hearing_ID=1565&Witness_ ID=4427. On corn ethanol, see also (14).

12. World Bank, *How the World Bank's Energy Framework Sells the Climate and Poor People Short* (World Bank, September 2006); www.nirs.org/ climate/background/energyreportfinal91806.pdf.

13. A. Elobeid, S. Tokgoz, *Removal of U.S. Ethanol Domestic and Trade Distortions: Impact on U.S. and Brazilian Ethanol Markets* (Working Paper 06-WP 427 October 2006); www.card.iastate.edu/publications/ DBS/PDFFiles/06wp427.pdf.

14. H. Shapouri, J. A. Duffield, M. S. Graboski, *Estimating the Net Energy Balance of Corn Ethanol* (U.S. Department of Agriculture, Economic Research Service, Office of Energy, Agricultural Economic Report No. 721); www.ers.usda.gov/publications/ aer721/AER721.PDF.

15. UNDP, UNDESA, WEC, *World Energy Assessment Overview 2004 Update* (UN Development Program, UN Department of Economic and Social Affairs, World Energy Council, 2005); www.undp.org/ energy/weaover2004.htm.

16. IEA, *Key World Energy Statistics* (International Energy Agency, 2006); www.iea.org/w/bookshop/add .aspx?id=144.

17. USDA, *The Economic Feasibility of Ethanol Production from Sugar in the United States* (U.S. Department of Agriculture, 2006).

18. I. C. Macedo, *Greenhouse Gas Emissions and Energy Balances in Bio-Ethanol Production and Use in Brazil;* www.unica.com.br/i_pages/files/gee3.pdf.

19. J. Woods, A. Bauen, *Technology Status Review and Carbon Abatement Potential of Renewable Transport Fuels in the UK* (UK Department of Transport and Industry Report B/U2/00785/REP URN 03/982); www.dti.gov.uk/files/file15003.pdf.

20. I thank O. Lucon and J. R. Moreira for useful discussions and contributions.

The Billion-Ton Biofuels Vision

CHRIS SOMERVILLE

Earth receives approximately 4000 times more energy from the Sun each year than humans are projected to use in 2050. Some of that energy can be captured through a variety of "renewable" sources, but the only form of solar energy harvesting that can contribute substantially to transportation fuel needs at costs competitive with fossil fuel is that captured by photosynthesis and stored in biomass. Brazil now obtains a quarter of its ground transportation fuel from ethanol produced by the fermentation of sugarcane sugar, and in the United States, approximately 90 corn grain-to-ethanol refineries produce about 4.5 billion gal of ethanol annually. The U.S. Energy Policy Act of 2005 would increase that production to 7.5 billion gal by 2012, but the United States

This article first appeared in *Science* (2 June 2006: Vol. 312, no. 5778). It has been revised for this edition.

currently uses about 140 billion gal of ground transportation fuel per year. To replace 30% of that amount with ethanol of equivalent energy content, as proposed recently by the Secretary of Energy, will require about 60 billion gal of ethanol. A recent analysis (1) concluded that the United States could produce about 1.3 billion dry tons of biomass each year in addition to present agricultural and forestry production. Because it is theoretically possible to obtain about 100 gal of ethanol from a ton of **cellulosic** biomass (such as corn stover, the stalks remaining after corn has been harvested), the United States could sustainably produce about 130 billion gal of fuel ethanol from biomass. In addition to a positive effect on the release of greenhouse gases, a biofuels program on this scale would have substantial economic and strategic advantages.

The creation of a new industry on that scale will require much basic and applied work on methods for converting plant **lignocellulose** to fuels, because several significant problems must be overcome to make the process ready for large-scale use. For example, cellulose is a recalcitrant substrate for **bioconversion**, and unacceptably large amounts of enzymes are required to produce sugar. Lignin occludes polysaccharides and inhibits **enzymatic hydrolysis** of these carbohydrates; energetically expensive and corrosive chemical pretreatments are required for its removal. The yeast currently used in large-scale ethanol production cannot efficiently ferment sugars other than glucose. And relatively low concentrations of ethanol kill microorganisms, requiring an expensive separation of the product from large volumes of yeast growth medium. These and other technical issues associated with this emerging industry have potential solutions, and many incremental advances can be envisioned. However, substantial public and private investment will be needed to meet the nation's goals. For instance, competitive funding for basic research in plant biology by all federal agencies totals only about 1% of the National Institutes of Health's budget. Small wonder that we do not know basic things such as the compo-

KEY TERMS

Cellulosic ethanol is a type of biofuel produced from lignocellulose, a structural material that comprises much of the mass of plants, which means that it makes more efficient use of the acreage required to grow the feedstocks than does using just part of a plant, such as the grain. Through bioconversion, cellulose is converted into chemicals such as liquid fuels. Enzymatic hydrolysis is the process by which enzymes (proteins that catalyze chemical reactions) break down cellulose into sugar, which can then be fermented into ethanol or other chemicals.

sition of the enzyme complex that synthesizes cellulose. Hopefully, a new U.S. Department of Energy report (2) that outlines the scientific issues will help set the direction for increased funding in this area. A national biofuels strategy will ultimately depend on massive support for basic curiosity-driven research in many aspects of nonmedical microbiology, plant biology, and chemical engineering. A fivefold increase in federal support during the next decade could readily be justified by the projected economic gains from the accelerated development of a cellulosic biofuel industry. To ensure parallel progress on the many different components of a biofuels strategy, it may be necessary to create a mission-oriented project similar to the Manhattan Project. Indeed, several of the national laboratories that were founded during the Manhattan era also pioneered some aspects of biofuel technology and could be a powerful source of relevant scientific and engineering expertise.

References and Notes

1. R. D. Perlack *et al.*, *Biomass as Feedstock for a Bioenergy and Bioproducts Industry: The Technical Feasibility of a Billion-Ton Annual Supply* (DOE/ GO-102005-2135, Oak Ridge National Laboratory, Oak Ridge, TN, 2005).

2. U.S. Department of Energy, *Breaking the Biological Barriers to Cellulosic Ethanol: A Joint Research Agenda* (U.S. DOE Office of Science and Office of Energy Efficiency and Renewable Energy, 2006); www.doegenomestolife.org/biofuels/.

The Biofuels Conundrum

DONALD KENNEDY

This story begins with good news, followed by a problem. Many governments around the world, and even some states within the United States, are finding ways to reduce greenhouse gas emissions. A major step is the almost completed buyout of the giant Texas electric utility TXU by an improbable concatenation of big investors, environmental organizations, and bankers. This promising deal would kill 8 of 11 projected coal-fired power plants and require the others to meet environmental performance standards. That's like a 15th seed making the Final Four or Watford winning the FA Cup. Meanwhile, there is hopeful talk in Silicon Valley about "clean tech," and "biofuels" is the new entrepreneurial mantra there. But the problem is that limiting

This article first appeared in *Science* (27 April 2007: Vol. 316, no. 5824). It has been revised for this edition.

carbon emissions with biofuels like ethanol is complex terrain, and most proposals turn out to carry external costs.

Let's start with the explosive growth of a corn ethanol industry in the tallgrass prairies of America's West. This boon for those rural economies succeeds a long history of dual-purpose farm legislation, in which production objectives are mixed with rural welfare goals. Refineries now number well over 100, with more being added rapidly, as farmers expand cultivation into lands formerly set aside for conservation and drop soybeans to make room for corn. Even if corn could yield 30% of the equivalent energy of gasoline (the goal set by the Secretary of Energy), that would create a whole array of collateral distortions. One would be its environmental impact in the United States. Another would be distortion of the price structure of an important grain commodity that is traded in world markets and used in livestock production. Will that make maize or meat more affordable

to poor countries that must import it, or to the poor people who need to buy it? Not likely.

Ethanol derived from sugarcane is better: Growing the plant is energetically less costly, and extraction and fermentation are more efficient. That's what must have interested President Bush during his "Chavez shadow tour" of South America in March. Of course, U.S. companies would love to import this valuable product, which now accounts for a quarter of the ground transportation fuel in Brazil. Despite such hopes, some senators supporting alcohol-from-corn have helped lay a heavy U.S. protective tariff on Brazilian alcohol derived from sugar. If we got rid of that, it would reduce total carbon emissions, though only if Brazil could expand its production substantially. Is there some deal in progress? Alas, nothing's up.

Sugar alcohol is better than corn alcohol, but palm oil is even better in your tank (though not in your martini). Its relatively high energy efficiency per unit volume makes it a good bio-diesel fuel. Trucks can run entirely on palm oil, although it is usually mixed with conventional fossil fuels. A large-scale effort is under way to convert lands in Indonesia to palm oil plantation agriculture, with plans to double current production in a few years. But again, the effort has a downside. Not only will the needed rainforest destruction (by burning) partly cancel any

energy advantage supplied by the palm oil, but the conversion will also threaten orangutans and other endangered species.

The best course is to abandon this cluttered arena and invest seriously in a direct approach. As Chris Somerville pointed out in this space (1), the conversion of cellulosic biomass (corn stover, wood chips) has a far higher potential for fuel production than any of the above biofuels. The challenge is biochemical: plant lignins occlude the cellulose cell walls, so they must be removed, and then the enzymology of cellulose conversion needs to be worked out. The technology is complex (2). No commercial reactor has yet been built, though six are funded. Some hope has been raised by new commitments, like the $500 million joint project between British Petroleum and the Universities of California and Illinois. Nevertheless, as Somerville notes, the sobering reality is that what the U.S. government spends on all of plant physiology is only one-hundredth of the research budget of the National Institutes of Health. That's far too little for a venture this important.

References and Notes

1. C. Somerville, *Science* 312, 1277 (2006).
2. R. F. Service, *Science* 315, 1488 (2007).

Carbon-Negative Biofuels from Low-Input High-Diversity Grassland Biomass

DAVID TILMAN, JASON HILL, CLARENCE LEHMAN

Biofuels derived from low-input high-diversity (LIHD) mixtures of native grassland perennials can provide more usable energy, greater greenhouse gas reductions, and less agrichemical pollution per hectare than can corn grain ethanol or soybean biodiesel. High-diversity grasslands had increasingly higher bioenergy yields that were 238% greater than monoculture yields after a decade. LIHD biofuels are carbon-negative because net ecosystem carbon dioxide (CO_2) sequestration [4.4 megagram (Mg) per hectare (ha) per year of CO_2 in soil and roots] exceeds fossil CO_2 release during biofuel production (0.32 Mg/ha per year). Moreover, LIHD biofuels can be produced on agriculturally degraded lands and thus need to neither displace food production nor cause loss of biodiversity via habitat destruction.

Globally escalating demands for both food (1) and energy (2) have raised concerns about the potential for food-based biofuels to be sustainable, abundant, and environmentally beneficial energy sources. Current biofuel production competes for fertile land with food production, increases pollution from fertilizers and pesticides, and threatens biodiversity when natural lands are converted to biofuel production. The two major classes of biomass for biofuel production recognized to date are monoculture crops grown on fertile soils (such as corn, soybeans, oilseed rape, switchgrass, sugarcane, willow, and hybrid poplar) (3–6) and waste biomass (such as straw, corn stover, and waste wood) (7–9). Here, we show the potential for a third major source of biofuel biomass: high-diversity mixtures of plants grown with low inputs on agriculturally degraded land, to address such concerns.

This article first appeared in *Science* (8 December 2006: Vol. 314, no. 5805). It has been revised for this edition.

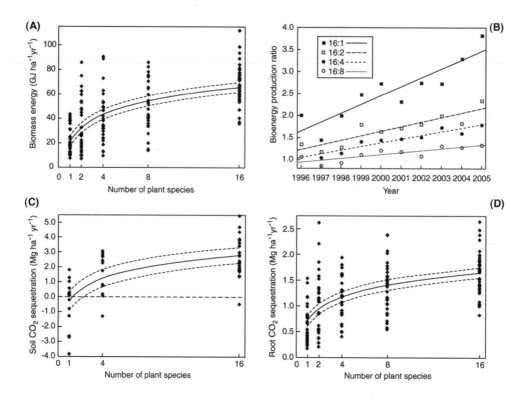

FIGURE 1. Effects of plant diversity on biomass energy yield and CO_2 sequestration for low-input perennial grasslands. (A) Gross energy content of harvested aboveground biomass (2003–2005 plot averages) increases with plant species number. (B) Ratio of mean biomass energy production of 16-species (LIHD) treatment to means of each lower-diversity treatment. Diverse plots became increasingly more productive over time. (C) Annual net increase in soil organic carbon (expressed as mass of CO_2 sequestered in upper 60 cm of soil) increases with plant diversity, as does (D) annual net sequestration of atmospheric carbon (as mass of CO_2) in roots of perennial plant species. Solid curved lines are log fits; dashed curved lines give 95% confidence intervals for these fits.

We performed an experiment on agriculturally degraded and abandoned nitrogen-poor sandy soil. We determined bioenergy production and ecosystem carbon sequestration in 152 plots, planted in 1994, containing various combinations of 1, 2, 4, 8, or 16 perennial herbaceous grassland species (table S1, online) (10). Species composition of each plot was determined by random draw from a pool of species. Plots were unfertilized, irrigated only during establishment, and otherwise grown with low inputs (10). The 16-species plots were the highest diversity, or the LIHD (low-input, high-diversity), treatment. All plots were burned in early spring to remove aboveground biomass

before growth began. Soil samples, collected before planting in 1994 and again in 2004, determined carbon sequestration in soil. Plots were sampled annually from 1996 to 2005 for aboveground biomass production.

Annual production of aboveground bioenergy (i.e., biomass yield multiplied by energy released upon combustion) (10) was an approximate log function of planted species number (Fig. 1A). On average for the last 3 years of the experiment (2003–2005), 2-, 4-, 8-, and 16-species plots produced 84%, 100%, 157%, and 238% more bioenergy, respectively, than did plots planted with single species. In a repeated-measures multivariate analysis of variance,

*Science*NOW Daily News, 7 January 2008

BIOFUELS ON A BIG SCALE

Robert F. Service

On paper, making biofuels from switchgrass and other perennials that need not be replanted seems like a no-brainer. Use the Sun's energy to grow the crop, and then convert it to liquid fuels to power our cars without the need for gasoline. But so far, experiments with these "cellulosic" crop-based fuels have only been conducted on small scales, leaving open the question of how feasible the strategy is. Now, the first large-scale study shows that switchgrass yields more than five times the energy needed to grow, harvest, and transport the grass and convert it to ethanol. The results could propel efforts to sow millions of hectares of marginal farmland with biofuel crops.

Previous studies on switchgrass plots suggested that ethanol made from the plant would yield anywhere from 343% to 700% of the energy put into growing the crop and processing it into biofuel. But these studies were based on lab-scale plots of about 5 square meters. So 6 years ago, Kenneth Vogel, a geneticist with the U.S. Department of Agriculture in Lincoln, Nebraska, and colleagues set out to enlist farmers for a much larger evaluation. Farmers planted switchgrass on 10 farms, each of which was between 3 and 9 ha. They then tracked the inputs they used—diesel for farm equipment and transporting the harvested grasses, for example—as well as the amount of grass they raised over a 5-year period. After crunching the numbers, Vogel and his colleagues found that ethanol produced from switchgrass yields 540% of the energy used to grow, harvest, and process it into ethanol. Equally important, the researchers found that the switchgrass is carbon-neutral, as it absorbs essentially the same amount of greenhouse gases while it's growing as it emits when burned as fuel.

A final significant finding, Vogel says, is that yields on farms using fertilizer and other inputs, such as herbicides and diesel fuel for farm machinery, were as much as six times higher than yields on farms that used little or no fertilizer, herbicides, or other inputs to grow a mixture of native prairie grasses. That result contrasts sharply with a controversial study published just over a year ago in *Science* that suggested that a mixture of prairie grasses farmed with little fertilizer or other inputs would produce a higher net energy yield than ethanol produced from corn (*Science*, 8 December 2006, p. 1598). Instead, the current study—published online in *Proceedings of the National Academy of Sciences*—shows that switchgrass farmed using conventional agricultural practices on less-than-prime cropland yields only slightly less ethanol per hectare on average than corn. "The bottom line is that low-input systems are not economically viable," Vogel says.

"This is a really important paper," says Christopher Somerville, who directs the Energy Biosciences Institute at the University of California, Berkeley. The impressive yield numbers, he adds, will likely serve as a baseline for future studies, because agricultural scientists are making rapid strides at creating new, higher-yielding switchgrass strains.

annual bioenergy production was positively dependent on the number of planted species ($F_{1, 155} = 68.4$, $P < 0.0001$), on time ($F_{9, 147} = 8.81$, $P < 0.0001$), and on a positive time-by-species number interaction ($F_{9, 147} = 11.3$, $P < 0.0001$). The interaction occurred because bioenergy production increased more through time in LIHD treatments than in monocultures and

Integrated gasification and combined cycle technology with Fischer-Tropsch hydrocarbon synthesis (IGCC-FT) combines several technologies to produce synthetic fuels and electricity. Fischer-Tropsch hydrocarbon synthesis is a catalyzed chemical reaction in which carbon monoxide and hydrogen are converted into liquid hydrocarbons that can be refined into synthetic fuels. Integrated gasification combined cycle, or IGCC, power plants convert biomass, coal, or petroleum residues into a synthetic gas, which is used to power a gas turbine generator. The waste heat from this generator is passed to a steam turbine system that generates additional electricity.

low-diversity treatments, as shown by the ratio of bioenergy in LIHD (16 species) plots to those in 8-, 4-, 2-, and 1-species plots (Fig. 1B).

The gross bioenergy yield from LIHD plots was 68.1 gigajoules (GJ)/ha per year. Fossil energy needed for biomass production, harvest, and transport to a biofuel production facility was estimated at 4.0 GJ/ha per year (table S2, online). Different biofuel production methods capture different proportions of bioenergy in deliverable, usable forms (Fig. 2) (10). Co-combustion of degraded-land LIHD biomass with coal in existing coal-fired electric generation facilities would provide a net gain of about 18.1 GJ/ha as electricity (11). Converting LIHD biomass into cellulosic ethanol and electricity is estimated to net 17.8 GJ/ha (12). Conversion into gasoline and diesel synfuels and electricity via integrated gasification and combined cycle technology with Fischer-Tropsch hydrocarbon synthesis (IGCC-FT) is estimated to net 28.4 GJ/ha (10, 13). In contrast, net energy gains from corn and soybeans from fertile agricultural soils are 18.8 GJ/ha for corn grain ethanol and 14.4 GJ/ha for soybean biodiesel (14). Thus, LIHD biomass converted via IGCC-FT yields 51% more usable energy per hectare from degraded infertile land than does corn grain ethanol from fertile soils. This higher net energy gain results from (i) low-energy inputs in LIHD biomass production, because the crop is perennial and

is not cultivated, treated with herbicides, or irrigated once established and likely requires only phosphorus replacement fertilization, because nitrogen is provided by legumes; (ii) the more than 200% higher bioenergy yield associated with high crop biodiversity; and (iii) the use of all aboveground biomass, rather than just seed, for energy. LIHD biofuels also provide much greater net energy outputs per unit of fossil fuel input than do current biofuels [see net energy balance (NEB) ratios of Fig. 2]. Fertile lands yield about 50% more LIHD biomass (and bioenergy) than our degraded soils (15, 16).

Annual carbon storage in soil was a log function of plant species number (Fig. 1C). For 1994–2004, there was no significant net sequestration of atmospheric CO_2 in monoculture plots [mean net release of CO_2 was 0.48 ± 0.44 Mg/ha per year (mean $\pm$ SE)], but, in LIHD plots, there was significant soil sequestration of CO_2 (2.7 ± 0.29 Mg/ha per year). Soil carbon storage occurred even though all aboveground biomass-based organic matter was removed annually via burning. Periodic resampling of soils in a series of prairie-like agriculturally degraded fields found carbon (C) storage rates similar to those of the LIHD treatment and suggested that this rate could be maintained for a century (17). The observed annual rate of change in soil C at a particular soil depth declined with depth ($P = 0.035$), suggesting that an additional

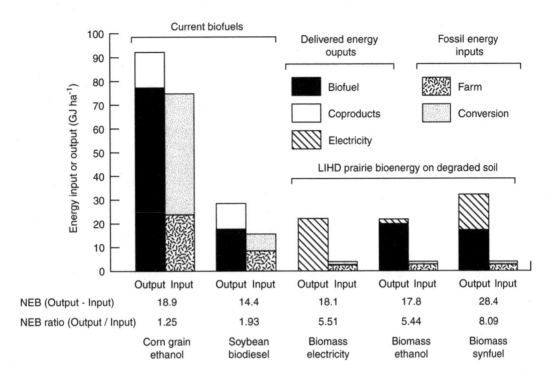

FIGURE 2. NEB for two food-based biofuels (current biofuels) grown on fertile soils and for LIHD biofuels from agriculturally degraded soil. NEB is the sum of all energy outputs (including co-products) minus the sum of fossil energy inputs. NEB ratio is the sum of energy outputs divided by the sum of fossil energy inputs. Estimates for corn grain ethanol and soybean biodiesel are from (14).

5% more C may be stored in soils deeper than we measured (below 60 cm depth).

In 2004, after 10 years of growth, atmospheric CO_2 sequestration in roots was a log function of plant species numbers (Fig. 1D). On an annual basis, 0.62 Mg/ha per year of atmospheric CO_2 was sequestered in roots of species grown in monocultures, and 160% more CO_2 (1.7 Mg/ha per year) was captured in roots of 16-species plots. Multiple regression showed that root CO_2 sequestration (Mg/ha of CO_2) increased as a log function of plant species number (S), as a log function of time ($Year$), and their interaction {$C_{root} = -1.47 + 6.16\log_{10}(S) + 9.64\log_{10}(Year) + 9.60[\log_{10}(S) - 0.613][\log_{10}(Year) - 0.782]$ where $Year = 3$ for 1997, the first time roots were sampled; overall $F_{3, 1260} = 191$, $P < 0.0001$; for $\log_{10}(S)$, $F_{1, 1260} = 398$, $P < 0.0001$; for $Year$, $F_{1, 1260} = 148$, $P = 0.0001$; for $S \times Year$, $F_{1, 1260} =$

27.3, $P = 0.0001$}. This regression suggests that most root carbon storage occurred in the first decade of growth; during the second decade, roots of 16-species plots are projected to store just 22% of C stored during the first decade. Measurements at greater depths in 10 LIHD plots suggest that 43% more C may be stored in roots between 30 and 100 cm.

LIHD plots had a total CO_2 sequestration rate of 4.4 Mg/ha per year in soil and roots during the decade of observation. Trends suggest that this rate might decline to about 3.3 Mg/ha per year during the second decade because of slower root mass accumulation. In contrast, the annual rate of CO_2 sequestration for monocultures was 0.14 Mg/ha per year for the first decade and projected to be indistinguishable from zero for subsequent decades.

Across their full life cycles, biofuels can be

carbon-neutral [no net effect on atmospheric CO_2 and other greenhouse gases (GHG)], carbon-negative (net reduction in GHG), or carbon sources (net increase in GHG), depending on both how much CO_2 and other greenhouse gases, expressed as CO_2 equivalents, are removed from or released into the atmosphere during crop growth and how much fossil CO_2 is released in biofuel production. Both corn ethanol and soybean biodiesel are net carbon sources but do have 12% and 41% lower net GHG emissions, respectively, than combustion of the gasoline and diesel they displace (14). In contrast, LIHD biofuels are carbon negative, leading to net sequestration of atmospheric CO_2 across the full life cycle of biofuel production and combustion (table S3, online). LIHD biomass removed and sequestered more atmospheric CO_2 than was released from fossil fuel combustion during agriculture, transportation, and processing (0.32 Mg/ha per year of CO_2), with net life cycle sequestration of 4.1 Mg/ha per year of CO_2 for the first decade and an estimated 2.7 to 3 Mg/ha per year for subsequent decades. GHG reductions from use of LIHD biofuels in lieu of gasoline and diesel fuel are from 6 to 16 times greater than those from use of corn grain ethanol and soybean biodiesel in lieu of fossil fuels (Fig. 3A).

LIHD biofuel production should be sustainable with low inputs of agrichemicals, as in our study. Legumes in LIHD plots can supply nitrogen (N) (18). In our experiment, total soil N of LIHD plots increased 24.5% ($P < 0.001$) from 1994–2004, but monoculture total soil N was unchanged ($P = 0.83$). However, some amount of N fertilization may be useful in dry habitats that lack efficient N-fixing species. Application of phosphorus or other nutrients may be needed if they are initially limiting or to replace nutrient exports (Fig. 3B). Production may be sustainable with low pesticide use, because plant disease incidence and invasion by exotic species are low in high-diversity plant mixtures (Fig. 3C) (19).

Switchgrass (*Panicum virgatum*), which is

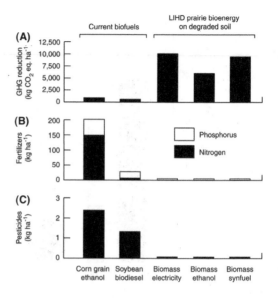

FIG. 3. Environmental effects of bioenergy sources. (A) GHG reduction for complete life cycles from biofuel production through combustion, representing reduction relative to emissions from combustion of fossil fuels for which biofuel substitutes. (B) Fertilizer and (C) pesticide application rates are U.S. averages for corn and soybeans (29). For LIHD biomass, application rates are based on analyses of table S2, online (10).

being developed as a perennial bioenergy crop, was included in our experiment. Switchgrass monocultures can be highly productive on fertile soils, especially with application of pesticides and fertilizer (20, 21). However, on our infertile soils, switchgrass monoculture bioenergy [23.0 ± 2.4 GJ/ha per year (mean ± SE)] was indistinguishable from mean bioenergy of monocultures of all other species (22.7 ± 2.7 GJ/ha per year) and yielded just a third of the energy of LIHD plots (10).

How much energy might LIHD biomass potentially provide? For a rough global estimate, consider that about 5×10^8 ha of agriculturally abandoned and degraded land producing biomass at 90 GJ/ha per year (22) could provide, via IGCC-FT, about 13% of global petroleum consumption for transportation and 19% of global electricity consumption (2). Without accounting for ecosystem CO_2 sequestration, this could eliminate 15% of current global CO_2 emissions, providing one of seven CO_2 reduction "wedges" needed to stabilize global CO_2 (23). GHG benefits would be larger if LIHD biofuels were, in general, carbon-negative, as might be expected if late-successional native plant species were used in LIHD biomass production on degraded soils [e.g., (17)].

The doubling of global demand for food and energy predicted for the coming 50 years (1, 2) and the accelerating use of food crops for biofuels have raised concerns about biodiversity loss if extant native ecosystems are converted to meet demand for both food and biofuels. There are also concerns about environmental impacts of agrichemical pollution from biofuel production and about climate change from fossil fuel combustion (14, 24–26). Because LIHD biomass can be produced on abandoned agricultural lands, LIHD biofuels need neither compete for fertile soils with food production nor encourage ecosystem destruction. LIHD biomass can produce carbon-negative biofuels and can reduce agrichemical use compared with food-based biofuels. Moreover, LIHD ecosystem management may provide other ecosystem services, including stable production of energy, renewal of soil fertility, cleaner ground and surface waters, wildlife habitat, and recreation (18, 19, 24, 27, 28). We suggest that the potential for biofuel production and carbon sequestration via low inputs and high plant diversity be explored more widely (30).

References and Notes

1. N. Fedoroff, J. Cohen, *Proc. Natl. Acad. Sci. USA* 96, 5903 (1999).
2. *International Energy Outlook* (DOE/EIA-0484, Energy Information Administration, U.S. Department of Energy, Washington, DC, 2006).
3. A. E. Farrell *et al.*, *Science* 311, 506 (2006).
4. J. Outlaw, K. Collins, J. Duffield, Eds., *Agriculture as a Producer and Consumer of Energy* (CABI, Wallingford, UK, 2005).
5. G. Keoleian, T. Volk, *Crit. Rev. Plant Sci.* 24, 385 (2005).
6. I. Lewandowski, J. Scurlock, E. Lindvall, M. Christou, *Biomass Bioenergy* 25, 335 (2003).
7. P. Gallagher *et al.*, *Environ. Resourc. Econ.* 24, 335 (2003).
8. Y. Zhang, M. Dubé, D. McLean, M. Kates, *Bioresource Technol.* 89, 1 (2003).
9. S. Kim, B. Dale, *Biomass Bioenergy* 26, 361 (2004).
10. Materials and methods are available as supporting material on *Science* Online.
11. M. Mann, P. Spath, *Clean Prod. Process.* 3, 81 (2001).
12. J. Sheehan *et al.*, *J. Ind. Ecol.* 7, 117 (2003).
13. C. Hamelinck, A. Faaij, H. den Uil, H. Boerrigter, *Energy* 29, 1743 (2004).
14. J. Hill, E. Nelson, D. Tilman, S. Polasky, D. Tiffany, *Proc. Natl. Acad. Sci. USA* 103, 11206 (2006).
15. P. Camill *et al.*, *Ecol. Appl.* 14, 1680 (2004).
16. C. Owensby, J. Ham, A. Knapp, L. Auen, *Global Change Biol.* 5, 497 (1999).
17. J. Knops, D. Tilman, *Ecology* 81, 88 (2000).
18. D. Tilman *et al.*, *Science* 294, 843 (2001).
19. J. Knops *et al.*, *Ecol. Lett.* 2, 286 (1999).
20. D. Parrish, J. Fike, *Crit. Rev. Plant Sci.* 24, 423 (2005).
21. K. Vogel, J. Brejda, D. Walters, D. Buxton, *Agron. J.* 94, 413 (2002).
22. M. Hoogwijk *et al.*, *Biomass Bioenergy* 25, 119 (2003).
23. S. Pacala, R. Socolow, *Science* 305, 968 (2004).
24. P. M. Vitousek, H. A. Mooney, J. Lubchenco, J. M. Melillo, *Science* 277, 494 (1997).
25. D. Tilman *et al.*, *Science* 292, 281 (2001).
26. G. Berndes, *Global Environ. Change* 12, 253 (2002).
27. J. A. Foley *et al.*, *Science* 309, 570 (2005).
28. D. Hooper *et al.*, *Ecol. Monogr.* 75, 3 (2005).
29. *Agricultural Chemical Usage 2004 and 2005 Field Crops Summaries* (National Agricultural Statistics Service, U.S. Department of Agriculture, Washington, DC, 2006).
30. Supported by grants from the University of Minnesota's Initiative for Renewable Energy and the Environment, the NSF (grant DEB 0080382), and the Bush Foundation. We thank S. Polasky, J. Fargione, E. Nelson, P. Spath, E. Larson, and R. Williams for comments.

Supporting Online Material

www.sciencemag.org/cgi/content/full/314/5805/1598/DC1
Materials and Methods
Tables S1 to S3
References

Commentary:
"Carbon-Negative Biofuels from Low-Input High-Diversity Grassland Biomass"

MICHAEL P. RUSSELLE, R. VANCE MOREY, JOHN M. BAKER,
PAUL M. PORTER, HANS-JOACHIM G. JUNG

Tilman *et al.* (Reports, 8 December 2006, p. 1598) argued that low-input high-diversity grasslands can provide a substantial proportion of global energy needs. We contend that their conclusions are not substantiated by their experimental protocol. The authors understated the management inputs required to establish prairies, extrapolated globally from site-specific results, and presented potentially misleading energy accounting.

Tilman *et al.* (1) reported that biofuels derived from diverse mixtures of native grassland perennials can provide greater energy yields and environmental benefits than monoculture grown on fertile soils. We agree that growing

herbaceous perennial species on land of marginal value for agriculture is desirable for several reasons, but we take issue with the authors' contention that low-input high-diversity (LIHD) prairie can provide a substantial contribution to our nation's energy needs. We argue that their experimental results do not substantiate their conclusions and that the authors overstated the global importance of their results.

Tilman *et al.* suggest that LIHD plantings could provide a sustainable source of harvestable biomass for fuel production, but they reported sample yields from an experiment in which nearly all the biomass was burned *in situ*, not harvested. Although several plant nutrients are lost from burned vegetation as gases or particulates, most cations are returned to the soil (2). With mechanical harvest, all nutrients are removed. Although legumes can replace nitrogen, nutrient replacement will be

This article first appeared in *Science* (15 June 2007: Vol. 316, no. 5831). It has been revised for this edition.

an important requirement for many marginal, and especially acidic, soils if yields are to be sustained. Limestone additions would be required to maintain symbiotic nitrogen fixation in soils with poor **pH buffering capacity**. Liming represents a major energy input (3, 4).

More seriously, the experimental approach of Tilman *et al.* is a form of double accounting with respect to carbon. The authors estimated harvestable biomass from small samples taken in late summer, then burned the remaining biomass on the plots the following spring [see supporting online material for (6)]. Combustion of this sort is incomplete, so some, if not most, of the soil carbon sequestration they measured is almost certainly due to charcoal additions that would not have occurred with harvest for biofuel production. Burning also has multiple, and often unpredictable, effects on prairie plant ecology. In general, burning reduces the presence of woody species in mixed stands, as the authors observed (1), but also helps control other undesirable species and may increase root biomass, tillering, soil temperature, and nitrification (2). With the exception of the decline in woody species, these benefits would not accrue with mechanical harvest of herbaceous perennials.

Tilman *et al.* (1) also ignored the difficulty of establishing and maintaining stands of native prairie species. Species composition was maintained artificially in the Cedar Creek plots with hand-weeding four times per year [see supporting online material for (5)], a practice that would be impossible in a commercial biomass production system. Because phenology differs among plant species, timing of biomass removal will influence species survival and composition of the grassland through interspecific competition. For instance, switchgrass, one of the dominant North American tallgrass prairie species, requires six weeks of regrowth to persist if harvested during the growing season (6). Resulting alterations in species dominance could affect grassland productivity and yield resilience under stress. Thus, the yields reported by Tilman *et al.* and their assumption of a 30-year useful stand

KEY TERM

pH buffering capacity is the ability of solid phase soil materials to resist changes in H+ ion concentration in the solution phase.

life may need to be reconsidered. In temperate climates, delaying harvest until after a killing frost in the fall would avoid the problem of interspecific competition during late summer regrowth, but it would also remove protective winter cover of great value for wildlife.

Tilman *et al.* base most of their report (1) on one experiment, yet extrapolate their results globally. The experiment was conducted at one site in central Minnesota, USA, on soils that have low soil–organic carbon, low water-holding capacity, and relatively shallow groundwater. The authors then estimated the amount of energy that might be provided by LIHD biomass, assuming 5×10^8 ha of "abandoned and degraded land." This land area (7) derives from studies estimating the potential for reforestation of degraded lands primarily in the tropics (8). However, we are not aware of large areas of "abandoned and degraded" agricultural lands in temperate regions of the globe that would permit establishment of large-scale LIHD biomass prairies without affecting food production, as the authors claim. In the entire United States, for example, there are only about 1.5×10^7 ha of land classified as idle cropland (9), and a substantial fraction of that area is in regions too arid to support the annual biomass yields projected in their report (1). We contend that, rather than attempt to make global calculations, the authors should have limited their interpretations to similar soil and climatic conditions in the United States, on clearly identified land where these practices could be implemented.

Finally, Tilman *et al.* make the misleading claim that LIHD biomass from degraded infertile land would produce more usable energy per hectare than corn grain ethanol from fertile soils [Fig. 2 in (1)]. The biofuel energy output (GJ/ha) for corn grain ethanol is four times as large as either of the two LIHD alternatives that include biofuel outputs. It also appears that most of the energy for the conversion process for LIHD biofuels, but not corn grain ethanol, was assumed to come from biomass co-products. Co-products from corn grain ethanol can provide all of the conversion energy (10), and applying them as conversion energy rather than co-product energy credit to their net energy balance ratios [Fig. 2 in (1)] results in a net energy of more than 50 GJ/ha for corn grain ethanol, with corresponding reductions in greenhouse gas emissions. Alternatively, using only half the corn stover produced from each hectare of corn grain that is used for ethanol production could provide all the energy required for distillation, or at least as much cellulosic ethanol as a hectare

of LIHD prairie, thereby substantially improving the energy balance of corn-based ethanol. To be meaningful, net energy and greenhouse gas emission comparisons among biofuel systems must be based on consistent assumptions about conversion technologies.

Alternative energy based on biomass has captured public attention, and considerable resources are being devoted to research, development, and implementation. There is potential for substantial environmental benefit but also for unproductive expenditure. Many agree that no single biomass feedstock or product will suffice because of the disparate economic, environmental, edaphic, climatic, technological, and logistical factors involved. We suggest that the results and conclusions presented by Tilman *et al.* be treated with appropriate caution until they have been subjected to more rigorous examination.

References and Notes

1. D. Tilman, J. Hill, C. Lehman, *Science* 314, 1598 (2006).

2. K. F. Higgins, A. D. Kruse, J. L. Piehl, *Effects of Fire in the Northern Great Plains* (South Dakota Extension Circular 761, 1989); www.npwrc.usgs .gov/resource/habitat/fire/index.htm (Version 16MAY2000).

3. M. S. Graboski, *Fossil Energy Use in the Manufacture of Ethanol* (National Corn Growers Association, St. Louis, MO, 2002).

4. H. Shapouri, J. A. Duffield, M. Wang, *The Energy Balance of Corn Ethanol: An Update* (USDA Office of the Chief Economist, Agric. Econ., Rep. no. 814, 2002).

5. D. Tilman *et al.*, *Science* 294, 843 (2001).

6. L. E. Moser, K. P. Vogel, in *Forages: An Introduction to Grassland Agriculture*, R. F. Barnes *et al.*, Eds. (Blackwell, Oxford, 1995), pp. 409–420.

7. M. Hoogwijk *et al.*, *Biomass Bioenergy* 25, 119 (2003).

8. A. Grainger, *Int. Tree Crops J.* 5, 31 (1988).

9. National Agricultural Statistics Service, *2002 Census of Agriculture*, Vol. 1, *State Level Data*. United States Table 8 (2004); www.nass.usda.gov/census/ census02/volume1/us/st99_1_008_008.pdf.

10. R. V. Morey, D. G. Tiffany, D. L. Hatfield, *Appl. Eng. Agric.* 22, 723 (2006).

Response to Commentary:
"Carbon-Negative Biofuels from Low-Input High-Diversity Grassland Biomass"

DAVID TILMAN, JASON HILL, CLARENCE LEHMAN

W e discovered that biofuels from low-input high-diversity (LIHD) mixtures of native perennial prairie plants grown on degraded soil can provide similar bioenergy gains and greater greenhouse gas benefits than current corn ethanol produced from crops grown in monoculture on fertile soil with high inputs. Russelle *et al.*'s technical concerns are refuted by a substantial body of research on prairie ecosystems and managed perennial grasslands.

Russelle *et al.* (1) raise several technical concerns that lead them to question our conclusions about the energetic and environmental advantages of biofuels derived from diverse mixtures of native perennial prairie plant species over biofuels from high-input annual food crops

This article first appeared in *Science* (15 June 2007: Vol. 316, no. 1567c). It has been revised for this edition.

such as corn (2). The nature of their comments suggests that research results well known in ecology may be less familiar to those outside the discipline. Indeed, our approach stands in marked contrast to that of conventional high-input agriculture. Each of their concerns, addressed below, is refuted by published studies of the ecology of high-diversity grasslands, and none of them has substantive effect on our original conclusions.

Russelle *et al.* (1) question the ability of LIHD prairie biomass to grow sustainably with low nutrient inputs. U.S. corn, in contrast, requires substantial inputs: 148 kg/ha of nitrogen (N), 23 kg/ha of phosphorus (P), and 50 kg/ha of potassium (K) annually (3). Leaching and erosional nutrient losses are much lower for perennial grasslands than for annually tilled row crops such as corn; hence, much lower inputs are needed. Moreover, we recommended harvesting prairie biomass when senescent in late autumn because this would "both yield greater biomass

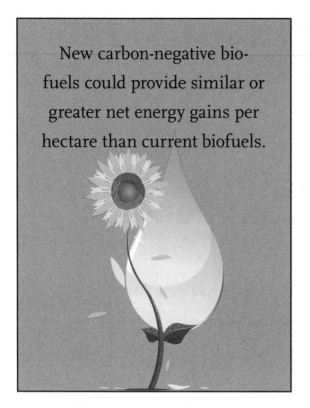

New carbon-negative bio-fuels could provide similar or greater net energy gains per hectare than current biofuels.

and decrease ecosystem loss of N, P, and other nutrients" [supporting online material for (4)]. Replacing nutrients removed by harvesting would require about 4 kg/ha of P and 6 kg/ha of K, should they be limiting (5, 6). LIHD mixtures needed no N fertilization because N fixation by legumes more than compensated for N exports in harvested biomass. Also, unlike some cultivated legumes, our native legumes grow well and fix N on acidic soils without liming (7). Moreover, several studies have shown that biomass yields of high-diversity grasslands are sustainable with low inputs. Annual hay yields from high-diversity Kansas prairie (8) showed no declines over 55 years despite no fertilization. Similarly, hay yields increased slightly during 150 years of twice-annual biomass removal in high-diversity unfertilized plots of the Park Grass experiment (9, 10). In total, nutrient inputs sufficient to sustain LIHD biomass production are an order of magnitude lower than for corn.

We showed that the dense root mass of LIHD prairie led to high rates of soil carbon sequestration (2). Russelle et al. (1) express concern that fire may have caused carbon storage through charcoal formation. However, published studies show that annual accumulation of charcoal carbon in frequently burned grasslands was < 1% of our observed rate of soil carbon accumulation (11, 12). Similarly, fire had no effect on soil black carbon levels in a six-year study of mixed-grass savanna (13). The concern about effects of late autumn mowing versus burning is also unfounded. Annual mowing and burning have similar effects on prairie biomass production (14, 15), and mowing does not cause prairie yields to decrease (8).

We proposed using mixtures of native prairie perennials for biofuels in part because, contrary to the assertion of Russelle et al. (1), such mixtures are easily established and require low or no inputs for maintenance. Indeed, prairie readily reestablishes itself from seed and displaces exotic plant species during natural succession on many degraded agricultural lands in the Great Plains (16). Prairie restoration, such as on the 6000 ha restored recently in Minnesota by The Nature Conservancy, is performed using agricultural machinery, not manual labor as Russelle et al. suggest. Our hand-weeding was done to maintain monoculture and low-diversity treatments. In contrast, the LIHD treatment led to rapid competitive displacement of exotic weedy and pasture species. LIHD plots were strikingly resistant to subsequent plant invasion and disease (17, 18). In portions of LIHD plots for which weeding had been stopped for three years, only 1.7% of total biomass came from invaders, which themselves were mainly native prairie perennials, and this invasion did not impact energy production.

Our one-sentence "rough global estimate" of the energy LIHD biomass might potentially provide was brief, but well supported and conservative. As to our estimated land base, 9×10^8 ha of global agricultural lands have been degraded so as to have "great reductions" in agricultural productivity, and an additional 3×10^8 ha are "severely degraded" and offer no agricultural utility (19, 20). A review of 17 studies found a median value of 7.1×10^8 ha of degraded land available globally for biofuel production (21). Our suggestion of 5×10^8 ha is 30% lower and is therefore a conservative estimate.

In our experiment (2), severely degraded land planted to LIHD mixtures had biomass production that was 46% as much as its native biome, temperate grassland (22). To be conservative, we assumed that LIHD mixtures of native species planted on degraded land would produce 20% less than we observed, that is, just 37% of the average of its native biome (22). Weighting this LIHD production estimate by the global area for each biome produced our estimate of 90 GJ/ha per year globally and our estimate of degraded lands potentially providing—through the integrated gasification combined cycle with Fischer-Tropsch (IGCC-FT) process—about one-seventh of the global transportation and electricity demand. We stand by that estimate. Further, we urge that the energy and carbon sequestration potential of low-input high-diversity mixtures of locally native plant species be explored for degraded lands around the world.

Our energy accounting was thorough and correct. We reported actual energy balances for U.S. corn ethanol and soybean biodiesel as currently produced (both of which cause net increases in greenhouse gases), and we compared them with three ways that LIHD prairie biomass might be used to produce carbon-negative biofuels (i.e., biofuels that, in total for their life cycle, decrease greenhouse gas levels). We showed that these new carbon-negative biofuels could provide similar or greater net energy gains per hectare than current biofuels.

The concerns of Russelle et al. (1) are refuted by a thorough consideration of the published literature. As to current biofuels, we agree that the energy and greenhouse gas benefits of corn ethanol could be improved, but we disagree about methods. First, burning the high-protein co-product of corn ethanol production to power ethanol production facilities, as Russelle et al. suggest, seems unwise because greater protein

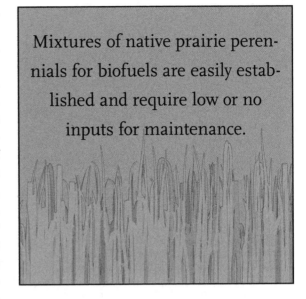

Mixtures of native prairie perennials for biofuels are easily established and require low or no inputs for maintenance.

production is required to meet global nutritional needs. Burning this protein is not an industry standard, nor is it discussed in any recent ethanol energy balance reviews (23, 24). Second, harvest and use of corn stover (the senescent stalks and leaves of corn plants) to power ethanol plants would likely cause soil–organic carbon levels to fall, and increase both carbon dioxide release and soil erosion. A better alternative would be powering corn ethanol plants with LIHD biomass, likely by gasification. If done properly, the ethanol produced could be carbon-neutral and have a markedly higher net energy gain than current corn ethanol.

The world's energy and climate problems are likely to be solved only by a combination of approaches and technologies, including wind and solar energy, increased energy efficiency, and renewable biofuels (25). Our research found that biofuels from LIHD biomass grown on degraded lands have substantial energy and greenhouse gas advantages over current U.S. biofuels. Moreover, LIHD production of renewable energy on agriculturally marginal lands could help ameliorate what might otherwise be an escalating conflict between food production, bioenergy production, and preservation of the world's remaining natural ecosystems. LIHD biofuels merit further exploration.

References and Notes

1. M. P. Russelle, R. V. Morey, J. M. Baker, P. M. Porter, H.-J. G. Jung, *Science* 316, 1567 (2007).

2. D. Tilman, J. Hill, C. Lehman, *Science* 314, 1598 (2006).

3. U.S. Department of Agriculture, *Agricultural Chemical Usage 2005 Field Crops Summary* (National Agricultural Statistics Service, USDA, Washington, DC, 2006).

4. D. Tilman, J. Hill, C. Lehman, *Science* 314, 1598 (2006).

5. M. R. Koelling, C. L. Kucera, *Ecology* 46, 529 (1965).

6. R. Samson *et al.*, *Crit. Rev. Plant Sci.* 24, 461 (2005).

7. D. Tilman, *Oikos* 58, 3 (1990).

8. J. Shortridge, *Geogr. Rev.* 63, 533 (1973).

9. D. Jenkinson *et al.*, *J. Agric. Sci.* 122, 365 (1994).

10. J. Silvertown, M. Dodd, K. McConway, J. Potts, M. Crawley, *Ecology* 75, 2430 (1994).

11. B. Glaser, W. Amelung, *Global Biogeochem. Cycles* 17, 1064 (2003).

12. X. Dai, T. Boutton, B. Glaser, R. Ansley, W. Zech, *Soil Biol. Biochem.* 37, 1879 (2005).

13. R. J. Ansley, T. W. Boutton, J. O. Skjemstad, *Global Biogeochem. Cycles* 20, GB3006 (2006).

14. L. Hulbert, *Ecology* 50, 874 (1969).

15. S. L. Collins, A. K. Knapp, J. M. Briggs, J. M. Blair, E. M. Steinauer, *Science* 280, 745 (1998).

16. R. S. Inouye *et al.*, *Ecology* 68, 12 (1987).

17. J. Knops *et al.*, *Ecol. Lett.* 2, 286 (1999).

18. J. Fargione, D. Tilman, *Ecol. Lett.* 8, 604 (2005).

19. L. R. Olderman, R. T. A. Hakkeling, W. G. Sombroek, *World Map of the Status of Human Induced Soil Degradation: An Explanatory Note, rev.* (International Soil Reference and Information Center, Wageningen, Netherlands, rev. ed. 2, 1990).

20. G. Daily, *Science* 269, 350 (1995).

21. G. Berndes, M. Hoogwijk, R. van den Broek, *Biomass Bioenergy* 25, 1 (2003).

22. W. H. Schlesinger, *Biogeochemistry: An Analysis of Global Change* (Academic Press, San Diego, ed. 2, 1997).

23. A. E. Farrell *et al.*, *Science* 311, 506 (2006).

24. R. Hammerschlag, *Environ. Sci. Technol.* 40, 1744 (2006).

25. S. Pacala, R. Socolow, *Science* 305, 968 (2004).

Can the Upstarts Top Silicon?

ROBERT F. SERVICE

Several nascent technologies are improving prospects for turning the Sun's rays into electricity. The success of any one of them could mean a big boost for solar power.

These are bright days for backers of solar power. The exuberance that previously pumped up dot-coms and biotech companies migrated in 2007 to solar energy, one of the hottest sectors in the emerging market for clean energy. Last year, solar energy companies around the globe hauled in nearly $12 billion from new stock offerings, loans, and venture capital funds. And although the markets have taken a bath in recent weeks due to investor fears about a coming recession in the United States, enthusiasm for solar's future remains strong. The industry

This article first appeared in *Science* (8 February 2008: Vol. 319, no. 5864). It has been revised for this edition.

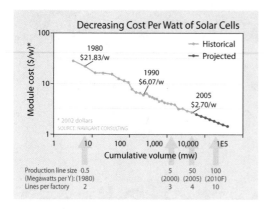

IMAGE 2. Heading for parity. Solar electricity still costs about five times as much as electricity from coal. But many experts expect economies of scale could close the gap by 2015. Source: Applied Materials. (http://www.sciencemag.org/cgi/content/full/319/5864/718/F4)

is growing at a whopping 40% a year. And the cost of solar power is dropping and expected to rival the cost of grid-powered electricity by the middle of the next decade (see figure above).

Still, there are clouds overhead. Solar power accounts for only a trivial fraction of the world's electricity. Silicon solar panels—which dominate the market with a 90% share—are already near their potential peak for converting solar energy to electricity and thus are unlikely to improve much more. A typical home's rooftop loaded with such cells can't produce enough power to meet the home's energy needs. That limitation increases the need for large-scale solar farms in sunny areas such as the American Southwest, which are far from large population centers. The bottom line is that the future of solar energy would be far brighter if researchers could make solar cells more efficient at converting sunlight to electricity, slash their cost, or both.

That's just what a new generation of solar

IMAGE 1 (PREVIOUS PAGE). Gold standard. Silicon solar cells dominate the market, but new competitors are rising fast. Courtesy of iStockphoto.

cell technologies aims to do. A raft of those technologies was on display here (1) late last year, as researchers reported how a broad array of recent advances in chemistry, materials science, and solid-state physics are breathing new life into the field of solar energy research. Those advances hold out the promise of solar cells with nearly double the efficiency of traditional silicon-based solar cells and of plastic versions that cost just a fraction of today's photovoltaics (PVs). "It's a really exciting time [in solar energy research]," says chemist David Ginger of the University of Washington, Seattle.

In the past few years, Ginger and others point out, solar researchers have hit upon several potential breakthrough technologies but have been stymied at turning that potential into solar cells able to beat out silicon. "The next couple of years will be important to see if we can overcome those hurdles," Ginger says. Although most of these novel cells are not yet close to commercialization, even one or two successes could dramatically change the landscape of worldwide energy production.

Minding the Gap

Beating silicon is a tall order. Although the top lab-based silicon cells now convert about 24% of the energy in sunlight into electricity, commercial cells still reach only 15% to 20%. In such traditional solar cells, photons hitting the silicon dump their energy into the semiconductor. That excites electrons, kicking them from their staid residence in the so-called valence band, where they are tightly bound to atoms, into the higher-energy conduction band, where they lead a more freewheeling existence, zipping through the material with ease. But if photons don't have enough energy to push electrons over this "band gap," the energy they carry is lost as heat. So is any energy the photons carry in excess of the band gap. Given the Sun's spectrum of rays and the fact that only certain red photons have the amount of energy that closely matches silicon's band gap, single silicon cells can convert at most

31% of the energy in sunlight into electricity—a boundary known as the **Shockley-Queisser limit**.

Engineers can boost the efficiency with a number of conventional strategies. One is to layer several light-absorbing materials that capture different portions of the solar spectrum—for example, by having one cell that absorbs mostly blue photons, while others absorb yellow and red photons. But such "tandem" cells are expensive to produce and thus are currently used primarily for high-end applications such as space flight.

But there may be other ways to capture more energy from the Sun. One strategy that has drawn a lot of attention in recent years is to find materials that generate multiple electronic charges each time they absorb a photon. Traditional silicon solar cells generate just one. In them, a layer of silicon is spiked with impurity atoms so that one side attracts negatively charged electrons while the other attracts positively charged electron vacancies, known as holes. Most light is absorbed near the junction between the two layers, creating electrons and holes that are immediately pulled in opposite directions.

In 1997, however, chemist Arthur Nozik of the National Renewable Energy Laboratory (NREL) in Golden, Colorado, and colleagues predicted that by using tiny nanosize semiconductor particles called quantum dots to keep those opposite charges initially very close

together, researchers could excite two or more electrons at a time. A paired electron and hole in close proximity, they reasoned, increases a quantum-mechanical property known as the **Coulomb interaction**. The greater this interaction, the more likely it is that an incoming energetic photon with at least twice the band-gap energy will create two electron-hole pairs with exactly the energy of the band gap—instead of one electron-hole pair with excess energy above the band gap, the other possible outcome that quantum mechanics allows. At least, that's how Nozik and his colleagues see it. Several other models of multiple-electron excitation exist, and theorists are still debating just what is behind the effect.

Four years ago, researchers led by Victor Klimov of Los Alamos National Laboratory in New Mexico reported the first spectroscopic evidence showing that multiple electron-hole pairs, known as excitons, were indeed generated in certain quantum dots. Nozik's team and others have since found the same effect in silicon and other types of quantum dots. And according to calculations by Nozik and NREL colleague Mark Hanna, the multiple-exciton generation (MEG)–based solar cells hit with unconcentrated sunlight have a maximum theoretical efficiency of 44%. Using special lenses and mirrors to concentrate the sunlight 500-fold, they

Science, Vol. 315, no. 5813, 9 February 2007

NORWAY: A NUCLEAR DEMONSTRATION PROJECT?

Daniel Clery

Egil Lillestøl is a man with a rather unusual mission: He wants his homeland of Norway to take the lead in developing a new form of nuclear power. Norway is Europe's largest petroleum exporter, from its North Sea oil and gas fields, and Lillestøl, a physicist at the University of Bergen, believes the country needs to do something about its carbon emissions. Norway has little experience with nuclear power but has one of the world's largest reserves of thorium. Lillestøl says Norway should pioneer a new, inherently safe form of nuclear reactor, called an energy amplifier, that runs on thorium. "It would be a good thing to have other [options] to stand on," Lillestøl says.

Carlo Rubbia, a Nobelist and former Director-General of Europe's particle physics lab CERN, championed the idea of the energy amplifier in the 1990s, and CERN researchers developed a design and tested some of the key ideas. A conventional fission reactor holds enough fissile material for a nuclear chain reaction to take place; neutron-absorbing rods ensure that the reaction doesn't run out of control, although this always remains a risk. The energy amplifier doesn't have enough fissile material to sustain a chain reaction. Instead, an accelerator fires high-energy particles into the fuel, prompting a cascade of fission reactions and producing heat. The amount of heat is proportional to the intensity of the beam, and the accelerator can be designed so that the amplifier can never overheat. Although the amount of waste produced is expected to be low, particle accelerators aren't cheap, and one with the necessary power has never been built.

Lillestøl wants Norway to pioneer this form of energy by funding and hosting a prototype—at a cost of about €550 million—and has made it a personal crusade to win over the Norwegian public and government. Lillestøl says he makes two or three presentations a week. Although the government is wary of nuclear power, after a debate in the national assembly, the energy minister called for an in-depth study. Norway is in a unique position to undertake such an enterprise, because it has been squirreling away oil revenue and has now amassed a fund of some $250 billion.

CERN's Jean-Pierre Revol, who worked on the energy amplifier at CERN, says that Lillestøl has made "a lot of political progress" in Norway. Renewed interest in nuclear power is generating curiosity about this technology, Revol says: "If it starts to fly, everyone will want to be part of it."

predicted, could boost the theoretical efficiency to about 80%—twice that of conventional cells hit with concentrated sunlight.

But reaching those higher efficiencies isn't easy. "One big hang-up is that no one has yet shown that you can extract those extra electrons," Nozik says. To harvest electricity, researchers must first break apart the pairs of electrons and holes, using an electric field across the cell to attract the opposite charges. That must happen fast, as electrons in excitons will collapse back into their holes within about 100 trillionths of a second if left side by side. If those charges can be separated, they must hop between successive quantum dots to find their way to an electrode, again without encountering an oppositely charged counterpart along the way. Unfortunately, the organic chemical coatings used to keep quantum dots stable and intact push the particles apart from one another, slowing down the charges.

Still, Nozik's group seems to be making progress. At the Materials Research Society meeting (1), Nozik reported preliminary results on solar cells made with arrays of lead selenide quantum dots. In such cells, a layer of quantum dots, and their organic coats, is spread between two electrodes. According to Nozik, spectroscopic studies indicate that two or three excitons are generated for every photon the dots absorb. And the researchers managed to separate the charges and get many of the electrons out, boosting the efficiency of the solar cells to about 2.5%, up from 1.62% of previous MEG-based cells.

To boost that efficiency further, Nozik says, one key will be to pack quantum dots closer together and in more regular arrays, making it easier for electronic charges to hop from one dot to the next to the electrodes where they are collected. Nozik's group is already experimenting with strategies for doing that, such as shrinking the organic groups that coat each dot and keep them separated from one another. Another needed improvement will be to find quantum dot materials better at generating multiple excitons. Nozik's lead selenide dots, for example,

must be hit with about 2.5 times the energy of a single excited electron to generate two excitons, meaning that extra energy is wasted. In the November 2007 issue of *Nano Letters*, however, Klimov and his colleagues reported that dots made from indium arsenide generate two excitons almost as soon as the energy of the incoming photons exceeds twice the band gap.

Other groups are hoping to use quantum dots as stepping-stones to cross the band gap in conventional semiconductor materials. The idea is to seed a semiconductor with an array of quantum dots, which will absorb photons that have too little energy to raise electrons above the band gap. The photons would excite electrons in the quantum dots to an intermediate level between the valence and conduction bands; then, a hit from a second low-energy photon would boost them the rest of the way into the conduction band.

Theoretical work by Antonio Luque of the Universidad Politécnica de Madrid in Spain suggests that such cells could achieve a maximum efficiency of 63% under concentrated sunlight. But here, too, the potential has been hard to realize. In practice, adding quantum dots to materials like an alloy of gallium arsenide seems to cause more losses than gains; the quantum

Science, Vol. 315, no. 5813, 792, 9 February 2007

PHOTOVOLTAICS IN FOCUS
John Bohannon

NEGEV DESERT, ISRAEL—
What looks like an upside-
down umbrella made of
mirrors is the future of
renewable energy—at least
according to its creator,
David Faiman, a physicist
who directs the Ben-
Gurion University National
Solar Energy Center here.
Photovoltaic cells have been
around for decades, but

COURTESY OF DAVID FAIMAN.

they've never been competitive with fossil fuels. Faiman claims to have found
a way to slash the price. "This technology is a real contender as a solution to
the world's energy problem," says physicist Robert McConnell of the National
Renewable Energy Laboratory in Golden, Colorado.

The secret ingredient is perched at the focus of his 10-ton reflector: a square
grid 10 cm across called a concentrator photovoltaic (CPV) cell. Instead of
spreading solar panels across a broad area to capture photons, Faiman uses
a reflector to concentrate the light 1000 times onto a small target. Traditional
silicon-based solar cells can't handle the heat, but a gallium arsenide–based
cell developed by a team at the Fraunhofer Institute for Solar Energy Systems in
Freiburg, Germany, "actually works far more efficiently at higher temperatures,"
says Faiman. By using a concentrating reflector, the system makes best use of
its most expensive component, the CPV cell, which Faiman estimates can reach
40% efficiency at converting sunlight to electricity. With this system, Faiman
believes he can build a power plant for less than the magic number of $1000
per kilowatt of electrical capacity. "Getting the price that low is feasible, but
only on a large scale," says McConnell, "and there's a long way to go from this
stage."

Large is exactly the scale Faiman is thinking along: spreading 20,000 CPV
cells over an area of 12 km^2 to generate 1 gigawatt. With mass-produced CPV
cells, Faiman estimates the cost at $1 billion. "Considering the savings, the sys-
tem can pay for itself within 2 decades," he says. The team is hoping to make it
happen sooner by increasing the efficiency of the CPV cell, for example, by adding extra layers of solar cells that capture a broader range of the wavelengths.

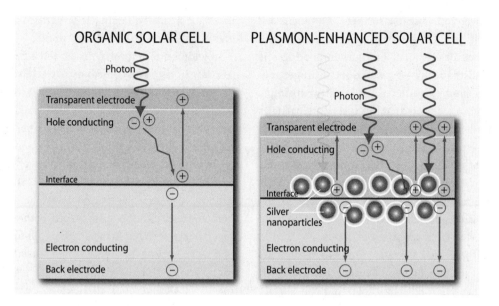

IMAGE 3. Better reception. In an organic solar cell, sunlight frees an electron (−) and an electron vacancy, or hole (+), which migrate to the border between different materials and then to oppositely charged electrodes (left). Adding metal nanoparticles (right) increases the light absorption and the number of charges generated. Adapted from M. Brongersma, P. Peumans, S. Fan.. Illustration from L. Creveling/Science.

dots also seem to attract electrons and holes and promote their recombination, thus losing the excess energy as heat.

Last year, Stephen Forrest and Guodan Wei, both of the University of Michigan, Ann Arbor, suggested a way around that problem: designing energetic barriers into their solar cells that discourage free charges from migrating to the quantum dots. At the meeting (1), Andrew Gordon Norman of NREL reported that his team has managed to grow such structures. The cells didn't outperform conventional gallium arsenide cells because too few quantum dots were packed into the structure to absorb enough low-energy photons to offset recombination losses. But Norman says he's working on solving that problem.

A Silver Lining

Many of the approaches to boosting the efficiency of solar cells require expensive materials or manufacturing techniques, so they are likely to increase capital costs. Some groups are exploring low-cost alternatives: light-absorbing plastics or other organic materials that can be processed without the expensive vacuum deposition machines most inorganics require. Unfortunately, organics waste much of the incoming light because they typically absorb only a relatively narrow range of frequencies in the solar spectrum. The key to boosting their efficiency, some groups believe, could be precious metals.

A layer of tiny silver or other metal nanoparticles added to a solar cell encourages an effect known as surface plasmon resonance, in which light triggers a collective excitation of electrons on the metal's surface. This causes the nanoparticles to act like antennas, capturing additional energy and funneling it to the active layer of the material to excite extra electrons (see figure above). At the meeting, electrical engineer Peter Peumans of Stanford University in Palo Alto, California, reported that when he and his colleagues added a layer of silver nanoparticles atop a conventional organic solar cell, they increased the efficiency of the device by 40%. Even though the overall efficiency of Peumans's devices is still dismally low—less than 1%—Ginger says the big jump in efficiency is "very promising."

Peumans notes that the silver nanoparticles work best when placed at the interface between two semiconducting layers in organic solar cells, one of which preferentially conducts electrons, the other, holes. In organic solar cells, excitons must migrate to just such an interface so that they can split into separate charges, which are then steered to opposite electrodes.

Other researchers have found in recent years that they can increase the efficiency of their organic solar cells by expanding the surface area of this interface. Instead of having flat layers lying atop one another like pages in a book, they create roughened layers that interpenetrate one another, a configuration known as a bulk heterojunction. Last year, researchers led by physicist Alan Heeger of the University of California, Santa Barbara, reported that they could use low-cost polymers to create tandem bulk heterojunction solar cells with an overall energy conversion efficiency of 6.5% (2). At the time, Heeger said that he expected that further improvements to the cells would propel them to market within three years.

Researchers are pursuing several strategies to improve these cells. One approach that may pay off down the road, Peumans says, is to incorporate metal nanoparticles into the random surface in these solar cells. That's not likely to be easy, he adds. But it may be possible to outfit the nanoparticles with chemical tethers that encourage them to bind to tags designed into the interface of the material. In theory, Peumans says, that would offer researchers the best of both worlds.

In addition to improving solar cell efficiencies, researchers and companies are also working on a host of technologies to make them cheaper. Nanosolar in San Jose, California, for example, has spent millions of dollars perfecting a new roll-to-roll manufacturing technology for making solar cells from thin films of copper indium gallium selenide atop a metal foil. Although they haven't reported the efficiency of their latest cells, they began marketing them in December 2007. Konarka, another roll-to-roll solar cell company in Lowell, Massachusetts, is working on a similar technology with plastic-based PVs. Other groups, meanwhile, are pushing the boundaries on everything from replacing quantum dots with nanowires that can steer excited charges more directly to the electrodes where they are harvested, to using modified ink-jet printers to spray films of quantum dots and other solar cell materials.

For now, there appears to be no shortage of ideas about creating new high-efficiency, low-cost cells. But whether any of these ideas will have what it takes to beat silicon and revolutionize the solar business remains the field's biggest unknown. "There are a lot of ways to beat the Shockley limit on paper, but it's difficult to realize in the real world," Nozik says. So far, it's not for want of trying.

References and Notes

1. Materials Research Society meeting, Boston, Massachusetts, 26-30 November 2007.
2. J. Y. Kim et al., Science 317, 5836 (2007).

What's Already Happened?

Introduction

DONALD KENNEDY

The debates about climate change have centered on a significant dimension: the steady increase, notably over the past 100 years, in the average temperature of the globe. Global warming has thus come to dominate the policy discussions that now surround the relationship between climate and energy. This issue has both a future and a past. The planet is already warmer on average by about 0.7°C. But much of the debate about future climate policy focuses on the challenge of predicting what will happen in the next 25, 50, or 100 years. That will be the subject of the next section of this book, but it is important to note here that much of the discussion about whether "global warming is real" concerns scientific disagreements about the general circulation models (GCMs) that are used in making predictions and their capacity for producing accurate forecasts. The authors of this book believe that the weight of the evidence favors the conclusions summarized in the most recent reports of the Intergovernmental Panel on Climate Change (IPCC). These project, first of all, that by mid-century, average global temperatures will have increased by between 2.5°C and 7.0°C, and sea level will have risen between about 20 and 70 cm. A combination of model prediction and the continuation of human activity in releasing carbon dioxide into the atmosphere—through the combustion of fossil fuels or the burning of forests—leads to those conclusions. There is some scientific disagreement here. It is believed by many scientists that the IPCC consensus underestimates the extent of sea level rise. An important issue here is whether the warming of Earth's climate will continue in a steady, ramp-like increase in average temperature, or whether the temperature change might pass the threshold of a dynamic, nonlinear process that could alter the climate dramatically. Were the major ice sheets in the Antarctic and Greenland to experience sudden increases in the rate of melting, that could lead to much larger increases in sea level, which could change the dynamics of ocean circulation—especially in the North Atlantic—and produce much colder temperatures in Europe.

The changes we have already observed, however, provide fewer sources for argument. Current models support that global warming will result in an increase of extreme weather events. Some of these have already occurred; for example, the intensity of hurricanes in both the Atlantic and Pacific have increased along with increases in sea surface temperature. And there are other impacts: droughts, of which there are historical records, are apt to be more severe and may result not only in public health hazards but also in an increase in the frequency of wildfires.

Fire, Drought, and Flood

This last connection is explored in the next section, in which a group of investigators associated with the Scripps Institution of Oceanography, the Laboratory of Tree-Ring Research at the University of Arizona, the University of California, and the U.S. Geological Survey have sought to analyze the history of wildfires in the western United States. In an analysis of fire history, their Research Article published in *Science* in 2006 shows an abrupt (fourfold) increase in wildfire frequency during the 1980s. Associated with global warming, higher springtime temperatures in the Rocky Mountains begin earlier than formerly, and of course snowmelt is accelerated. The result is drier forests and a fire season that begins much earlier and lasts longer. The increase in early snowmelt has been associated not only with an increased fire frequency, but also with increases in the area burned.

The focus on snow and snowmelt naturally calls to mind the question of how much effect global warming may have on the amount of precipitation and on its timing—and the importance of these changes for human residents. Higher average temperatures not only influence the melting times of mountain snowfall. They also strongly influence the water content of the snowpack that accumulates during the winter. This is a critically important issue, since melting snowpacks supply water for irrigation, industrial use, and human consumption on dependent flatlands. Thus, for example, residents of coastal California and the Central Valley cast anxious glances toward the High Sierra hoping for a good snowpack and listen eagerly to the regular forecasts of its water content.

Of course, precipitation is not all in the form of snow. Rainfall and its distribution is critical to human life: an excess produces floods and landslides, whereas an insufficiency may result in droughts. Wentz, Ricciardulli, Hilburn, and Mears analyzed the question "How Much More Rain Will Global Warming Bring?" in a 2007 *Science* Report. It is known that evaporation of water from the ocean surface increases as the average temperature increases, which in turn should make precipitation events more frequent and severe, at least in certain regions. But various models predict that rainfall would increase far more slowly than the data on evaporation rates would suggest. So Wentz *et al.* compared results from different models and looked at historical data on the increase in global rainfall to see whether the models used to "hindcast" the observed changes in precipitation did so accurately. The empirical observations actually showed that the increase in rainfall—over a time period that included two El Niño events—matched the increase in evaporation much more closely than the models had predicted. The result clearly suggests that as global warming continues, we can expect the kinds of rainfall events that have already been experienced as hazardous landslides and local flooding in some parts of the world—and droughts in other places.

One area that's already suffering from lack of water is the Gaza Strip—called a "worst-case scenario for water-resource planners" by John Bohannon. In his article, "Running Out of Water—and Time," Bohannon shows how geography, politics, and war combine to make clean drinking water an unaffordable luxury for many of Gaza's 1.4 million residents. While Gaza is unique in its particulars, the tangled relationship between natural resources and political strife is unfortunately not. As water grows scarcer in many parts of the world, cooperation instead of conflict will become ever more critical.

Impacts on Living Things

Among the biological consequences of global warming that might be expected are those that depend on climate cues that a variety of organisms take from annual temperature cycles. The flowering times of plants, the arrival times of migrating birds, and collateral events are collectively called phenological effects. In 2006, Jonzén *et al.*—an international group led from Sweden but including representation from Italy and other Scandinavian countries—published a *Science* Report examining what has happened to the arrival times in Europe of birds that migrate over various distances. Making use of sites in Scandinavia and one located in southern Italy, the authors compared arrival times of short-distance migrants (those whose migrations begin from points close to the breeding grounds) with those of long-distance migrants whose trips begin in Southern Africa. One theory predicted that short-distance migrants would be most affected by global warming; the rationale is that because the climate signals where they begin their migration closely resemble those on the breeding grounds, birds might advance their departure on that basis. But the data showed that long-distance migrants advanced their arrival times more—especially in the early phases of the migration. Different hypotheses might explain the effect: either these migrants accelerated their speed during the northward trip, or they left earlier, or both. The evidence suggested that the effects are not due to changes in forage availability on the wintering grounds, which might have triggered an earlier departure. Instead, it seems more probable that rapid evolutionary change, brought about during the period in which global warming was changing the signal, was responsible.

Many animals exhibit strong temperature optima and exhibit preferences dependent on them. For stationary species, global warming would be expected to initiate poleward or up-slope changes in distribution. For migrants, the situation is more complex. Changes in the distribution of prey species or food plants could take place both on the wintering ground and in the breeding locale. Either could result in evolutionary changes in behavior that would be reflected in altered migratory patterns.

Perhaps the most dramatic indicators of the influence of global climate change on animal species are the sad pictures of polar bears, apparently trapped on ice floes floating in the Arctic Ocean. Climate models have consistently shown that the effects of rising average global temperatures will be more severe at high latitudes, and popular recent accounts have suggested that the "Northwest Passage," once a fiction, may become open to shipping.

The Feedback Effect

In "Perspectives on the Arctic's Shrinking Sea-Ice Cover," Mark Serreze, Marika Holland, and Julienne Stroeve—from the University of Colorado and the National Center for Atmospheric Research, both in Boulder—have examined the trends in Arctic sea-ice extent since 1979. Much of this work has been based on satellite images and derived microwave data. The data show that average ice extent in any given month has declined in the period 1979–2006—with the decrease especially marked for September, the month in which the ice cover is typically minimal. For that month, the ice loss between 1979 and 2000 amounted to an area the size of Alaska—and the 2005 ice extent was the lowest in 50 years, according to satellite data and a variety of ground-level reports.

Ice loss is not only affected by local surface warming; it is also influenced by ocean currents and ocean-atmosphere coupling that may vary from year to year. Despite these factors, the accumulation of greenhouse gases plays a very significant role, and the models used by the IPCC did reasonably well at describing the changes that had occurred during the 1979–2006 period.

What does this tell us about possible futures? An especially important aspect of the loss of Arctic ice is what is called the ice-albedo feedback cycle. An ice surface reflects almost all impinging radiation, whereas the dark surface left after the ice melts absorbs light, including the long-wavelength radiation that warms the surface layers. That will, in turn, delay further ice formation as winter approaches and will contribute to thinner ice in the following spring. Thus the effect of global warming at high latitudes exerts an additional influence on the climate balance.

Interest in high-latitude ice is not restricted to what floats on the sea surface. Antarctica and Greenland are covered with massive ice caps, which have attracted attention because of their potential influence on global sea levels—which would rise by about 70 m if all their ice were to melt. The mass of a large glacier is determined by the amount of snow added by precipitation, which may be expected to rise as global warming takes place. This input of new ice is balanced by loss through melting, which will also increase its rate as the temperature is increased. If these two are in equilibrium, the ice cap or glacier is said to be in mass balance. Measurements of the Greenland Ice Cap tend to show a loss—but past estimates for Antarctica have given different answers to the question of whether that ice cap is losing or gaining mass.

Andrew Shepherd and Duncan Wingham, experts from two Centres for Polar Observation and Modelling, report what has more recently been learned from satellite-based measurements of the mass imbalance in Antarctica and the Greenland Ice Cap. Some of these are estimates of the gravitational attraction of the ice sheets; others are altimeter or interferometry measurements that can be used to calculate mass. The methods do not always agree fully. The authors are nevertheless able to reach some conclusions about the contribution of losses in mass balance to global sea level rise. This turns out to be rather small, accounting for only a little more than one-tenth of the observed annual sea level gain of 3.0 mm per year. That is a very small contribution compared with estimates of much larger increases that might occur if significant parts of either ice mass were to melt because of a dramatic change.

An especially interesting question is raised by the observation that the melting of many glaciers is accelerating in the most recent years: more streams are occurring on the surface, and meltwater may be reaching the glacier's bed and accelerating movement. If the very recent changes forecast a continuing acceleration of ice lost, then past estimates of sea level rise may have to be revised upward.

Warming and Earlier Spring Increase Western U.S. Forest Wildfire Activity

A. L. WESTERLING, H. G. HIDALGO, D. R. CAYAN, T. W. SWETNAM

W estern United States forest wildfire activity is widely thought to have increased in recent decades, yet neither the extent of recent changes nor the degree to which climate may be driving regional changes in wildfire has been systematically documented. Much of the public and scientific discussion of changes in western United States wildfire has focused instead on the effects of 19th- and 20th-century land-use history. We compiled a comprehensive database of large wildfires in western United States forests since 1970 and compared it with hydroclimatic and land surface data. Here, we show that large wildfire activity increased suddenly and markedly in the mid-1980s, with higher large-wild-

This article first appeared in *Science* (originally published in *Science* Express on 6 July 2006, *Science* 18 August 2006: Vol. 313, no. 5789). It has been revised for this edition.

fire frequency, longer wildfire durations, and longer wildfire seasons. The greatest increases occurred in mid-elevation, Northern Rockies forests, where land-use histories have relatively little effect on fire risks, and are strongly associated with increased spring and summer temperatures and an earlier spring snowmelt.

Wildfires have consumed increasing areas of western U.S. forests in recent years, and fire-fighting expenditures by federal land-management agencies now regularly exceed $1 billion/year (1). Hundreds of homes are burned annually by wildfires, and damages to natural resources are sometimes extreme and irreversible. Media reports of recent, very large wildfires [> 100,000 hectare (ha)] burning in western forests have garnered widespread public attention, and a recurrent perception of crisis has galvanized legislative and administrative action (1–3).

Extensive discussions within the fire management and scientific communities and the media seek to explain these phenomena, focusing on either land-use history or climate as primary causes. If increased wildfire risks are driven primarily by land-use history, then ecological restoration and fuels management are potential solutions. However, if increased risks are largely due to changes in climate during recent decades, then restoration and fuels treatments may be relatively ineffective in reversing current wildfire trends (4, 5). We investigated 34 years of western U.S. (hereafter, "western") wildfire history together with hydroclimatic data to determine where the largest increases in wildfires have occurred and to evaluate how recent climatic trends may have been important causal factors.

Competing Explanations: Climate versus Management

Land-use explanations for increased western wildfire note that extensive livestock grazing and increasingly effective fire suppression began in the late 19th and early 20th centuries, reducing the frequency of large surface fires (6–8). Forest regrowth after extensive logging beginning in the late 19th century, combined with an absence of extensive fires, promoted forest structure changes and biomass accumulation, which now reduce the effectiveness of fire suppression and increase the size of wildfires and total area burned (3, 5, 9). The effects of land-use history on forest structure and biomass accumulation are, however, highly dependent upon the "natural fire regime" for any particular forest type. For example, the effects of fire exclusion are thought to be profound in forests that previously sustained frequent, low-intensity surface fires [such as Southwestern ponderosa pine and Sierra Nevada mixed conifer (2, 3, 10, 11)] but of little or no consequence in forests that previously sustained only very infrequent, high-severity crown fires [such as Northern Rockies lodgepole pine or spruce-fir (1, 5, 12)].

In contrast, climatic explanations posit that increasing variability in moisture condi-

tions (wet/dry oscillations promoting biomass growth, then burning) and/or a trend of increasing drought frequency and/or warming temperatures have led to increased wildfire activity (13, 14). Documentary records and proxy reconstructions (primarily from tree rings) of fire history and climate provide evidence that western forest wildfire risks are strongly positively associated with drought concurrent with the summer fire season and (particularly in ponderosa pine–dominant forests) positively associated to a lesser extent with moist conditions in antecedent years (13–18). Variability in western climate related to the Pacific Decadal Oscillation and intense El Niño/La Niña events in recent decades, along with severe droughts in 2000 and 2002, may have promoted greater forest wildfire risks in areas such as the Southwest, where precipitation anomalies are significantly influenced by patterns in Pacific sea surface temperature (19–22). Although corresponding decadal-scale variations and trends in climate and wildfire have been identified in paleo studies, there is a paucity of evidence for such associations in the 20th century.

We describe land-use history versus climate as competing explanations, but they may be complementary in some ways. In some forest types, past land uses have probably increased the sensitivity of current forest wildfire regimes to climatic variability through effects on the

What's Already Happened?

*Science*NOW Daily News, 29 April 2008

FIRE AND BRIMSTONE, CRETACEOUS STYLE

Phil Berardelli

Talk about an oil crisis. An international team examining sediments from the end of the dinosaur age has discovered microscopic carbon spheres that can be produced only from burning fossil fuels. If confirmed, the finding means that the dinosaurs might have been wiped out, at least partly, by an oil-fueled conflagration.

Two major lines of evidence implicate an extraterrestrial object as the dinosaurs' killer. A thin layer of the mineral iridium in sediments deposited about 65 million years ago points to a catastrophic impact. Scientists have also located the probable site of the strike: the Chicxulub Crater on the Yucatan Peninsula in Mexico.

But the impact might not have been big enough to exterminate the dinosaurs—something else was needed. Further research by several groups provided a possibility: massive forest fires ignited by the heat of the blast could have pumped enough carbon dioxide into the atmosphere to cause a period of runaway global warming, cooking the dinosaurs. Or the fires might have spewed enough soot to block out the sun and kill off the plants on which herbivorous dinos fed, an effect that would have been felt all the way up the dino food chain. The problem with the fire scenario is that although scientists have found soot at the Cretaceous-Tertiary boundary—also called the K-T boundary—from the time of the dinosaurs' demise, they have only unearthed a few traces of charred plant remains.

The new potential explanation derives from another component of the soot in the K-T sediments: distinctive carbon globs known as cenospheres, which form only when fossil hydrocarbons, such as coal and crude oil, burn. The team reported in *Geology* that it found cenospheres at 8 of 13 sites it examined around the world, and the objects were present only at the K-T boundary, not above or below it. The researchers suspect the Chicxulub object plowed into a huge oil reservoir in the Gulf of Mexico, like the ones feeding offshore platforms there today. The impact first vaporized the oil and then ignited it in the atmosphere, causing an enormous, spreading fireball probably hundreds of kilometers wide. Whether the conflagration was enough to do in the dinos—via fire, soot, or global warming—remains unknown, but it would have spared a variety of critters, including the ancestors of today's mammals.

The paper is an "eye opener," says paleobotanist Peter Wilf of Pennsylvania State University in State College. It makes "a strong case for the true source of the mysterious soot" in the K-T layer and casts doubt "on the venerable global wildfire hypothesis," he says. The results should help researchers "better understand environmental conditions during the disaster and are thus very important for interpreting the mass extinction and life's subsequent recovery."

quantity, arrangement, and continuity of fuels. Hence, an increased incidence of large, high-severity fires may be due to a combination of extreme droughts and overabundant fuels in some forests. Climate, however, may still be the primary driver of forest wildfire risks on inter-annual to decadal scales. On decadal scales, climatic means and variability shape the character of the vegetation [e.g., species populations and their drought tolerance (23) and biomass (fuel) continuity (24), thus also affecting fire regime responses to shorter-term climate variability]. On interannual and shorter time scales, climate variability affects the flammability of live and dead forest vegetation (13–19, 25).

High-quality time series are essential for evaluating wildfire risks, but for various reasons (26), previous works have not rigorously documented changes in large-wildfire frequency for western forests. Likewise, detailed fire-climate analyses for the region have not been conducted to evaluate what hydroclimatic variations may be associated with recent increased wildfire activity, and the spatial variations in these patterns.

We compiled a comprehensive time series of 1166 large (> 400 ha) forest wildfires for 1970 to 2003 from federal land-management units containing 61% of western forested areas (and 80% above 1370 m) (26) (see fig. S1 online). We compared these data with corresponding hydro-climatic and land surface variables (26–34) to address where and why the frequency of large forest wildfire has changed.

Increased Forest Wildfire Activity

We found that the incidence of large wildfires in western forests increased in the mid-1980s (Fig. 1) [hereafter, "wildfires" refers to large-fire events (> 400 ha) within forested areas only (26)]. Subsequently, wildfire frequency was nearly four times the average of 1970 to 1986, and the total area burned by these fires was more than six and a half times its previous level. Interannual variability in wildfire frequency is strongly associated with regional spring and summer temperature (Spearman's correlation of 0.76, $P < 0.001$, $n = 34$). A second-order polynomial fit to the regional temperature signal alone explains 66% of the variance in the annual incidence of these fires, with many more wildfires burning in hotter than in cooler years.

The length of the wildfire season also increased in the 1980s (Fig. 1). The average season length (the time between the reported first wildfire discovery date and the last wildfire control date) increased by 78 days (64%), comparing 1970–1986 with 1987–2003. Roughly half of that increase was due to earlier ignitions, and half, to later control (48% versus 52%, respectively). Later control dates were no doubt partly

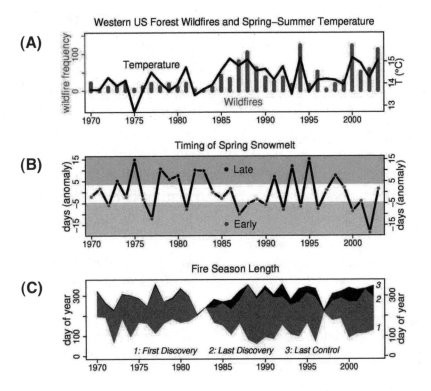

FIG. 1. (A) Annual frequency of large (> 400 ha) western U.S. forest wildfires (bars) and mean March-through-August temperature for the western United States (line) (26, 30). Spearman's rank correlation between the two series is 0.76 (P < 0.001). Wilcoxon test for change in mean large–forest fire frequency after 1987 was significant (W = 42; P < 0.001). (B) First principle component of center timing of streamflow in snowmelt-dominated streams (line). Low (blue shading), middle (no shading), and high (gray shading) tercile values indicate early, mid, and late timing of spring snowmelt, respectively. (C) Annual time between first and last large-fire ignition, and last large-fire control.

due to later ignition dates, given that the date of the last reported wildfire ignition increased by 15 days, but a substantial increase in the length of time the average wildfire burned also played a role. The average time between discovery and control for a wildfire increased from 7.5 days for 1970–1986 to 37.1 days for 1987–2003. The annual length of the fire season and the average time each fire burned were also moderately correlated with the regional spring and summer temperatures [Spearman's correlations of 0.61 (P < 0.001) and 0.55 (P < 0.001), respectively].

The greatest increase in wildfire frequency has been in the Northern Rockies, which account for 60% of the increase in large fires. Much of

the remaining increase (18%) occurred in the Sierra Nevada, Southern Cascades, and Coast Ranges of Northern California and southern Oregon (see "Northern California" in fig. S2 online). The Pacific Southwest; the Southern Rockies; the Northwest; coastal, central, and southern California; and the Black Hills account for 11%, 5%, 5%, < 1%, and < 1%, respectively. Interestingly, the Northern Rockies and the Southwest show the same trend in wildfire frequency relative to their respective forested areas. However, the Southwest's absolute contribution to the western regional total is limited by its smaller forested area relative to higher latitudes.

Increased wildfire frequency since the mid-1980s has been concentrated between 1680 and 2590 m in elevation, with the greatest increase centered around 2130 m. Wildfire activity at these elevations has been episodic, coming in pulses during warm years, with relatively little activity in cool years, and is strongly associated with changes in spring snowmelt timing, which in turn is sensitive to changes in temperature.

Fire Activity and the Timing of the Spring Snowmelt

As a proxy for the timing of the spring snowmelt, we used Stewart and colleagues' dates of the center of mass of annual flow (CT) for snowmelt-dominated streamflow gauge records in western North America (32–34). The annual wildfire frequency for the region is highly correlated (inversely) with CT at gauges across the U.S. Pacific Northwest and interior West, indicating a coherent regional signal of wildfire sensitivity to snowmelt timing. The negative sign of these correlations indicates that earlier snowmelt dates correspond to increased wildfire frequency. Following Stewart et al., we used the first principal component (CT_1) of CT at western U.S. streamflow gauges as a regional proxy for interannual variability in the arrival of the spring snowmelt (Fig. 1) (26, 32). This signal had its greatest impact on wildfire frequency between elevations of 1680 and 2590 m, with a nonlinear response at these elevations to variability in snowmelt timing. Overall, 56% of wildfires and 72% of area burned in wildfires occurred in early (i.e., lower tercile CT_1) snowmelt years, whereas only 11% of wildfires and 4% of area burned occurred in late (i.e., upper tercile CT_1) snowmelt years.

Temperature affects summer drought, and thus flammability of live and dead fuels in forests through its effect on evapotranspiration and, at higher elevations, on snow. Additionally, warm spring and summer temperatures were strongly associated with reduced winter precipitation over much of the western United States

KEY TERM

Center of mass of annual flow: a date such that half the mass of water that flows by a given location in a stream in a particular year flows by prior to that date and half flows by after that date. Typically, in the western United States this calculation is for a "water-year" that starts on October 1 and goes to September 30, in order to capture the whole winter, when most of the precipitation occurs in the region.

(Fig. 2). The arrival of spring snowmelt in the mountains of the western United States, represented here by CT_1, is strongly associated with spring temperature (26). Average spring and summer temperatures throughout the entire region are significantly higher in early than in late years (Fig. 2), peaking in April. The average difference between early and late April mean monthly temperatures in forested areas was just over 2°C, and it increased with elevation.

Snow carries over a substantial portion of the winter precipitation that falls in western mountains, releasing it more gradually in late spring and early summer, providing an important contribution to spring and summer soil moisture (35). An earlier snowmelt can lead to an earlier, longer dry season, providing greater opportunities for large fires due both to the longer period in which ignitions could potentially occur and to the greater drying of soils and vegetation. Consequently, it is not surprising that the incidence of wildfires is strongly associated with snowmelt timing.

Changes in spring and summer temperatures associated with an early spring snowmelt come in the context of a marked trend over the period of analysis. Regionally averaged spring and summer temperatures for 1987 to

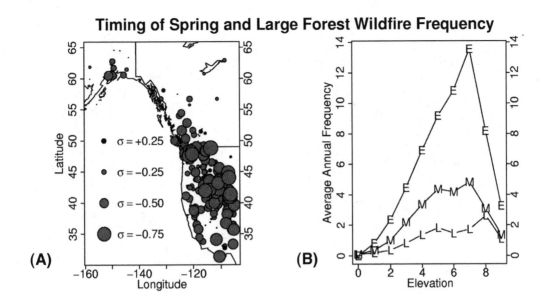

Timing of Spring and Large Forest Wildfire Frequency

(A)

(B)

FIGURE 2. (A) Pearson's rank correlation between annual western U.S. large (>400 ha) forest wildfire frequency and streamflow center timing. *x* axis, longitude; *y* axis, latitude. (B) Average frequency of western U.S. forest wildfire by elevation and early, mid-, and late snowmelt years from 1970 to 2002. See Fig. 1B for a definition of early, mid-, and late snowmelt years.

2003 were 0.87°C higher than those for 1970 to 1986. Spring and summer temperatures for 1987 to 2003 were the warmest since the start of the record in 1895, with 6 years in the 90th percentile—the most for any 17-year period since the start of the record in 1895 through 2003—whereas only 1 year in the preceding 17 years ranked in the 90th percentile. Likewise, 73% of early years since 1970 occurred in 1987 to 2003 (Fig. 1).

Spatial Variability in the Wildfire Response to an Earlier Spring

Vulnerability of western U.S. forests to more frequent wildfires due to warmer temperatures is a function of the spatial distribution of forest area and the sensitivity of the local water balance to changes in the timing of spring. We measured this sensitivity using the October-to-September moisture deficit—the cumulative difference between the potential evapotranspiration due to temperature and the actual evapotranspiration constrained by available

moisture—which is an important indicator of drought stress in plants (24). We used the percentage difference in the moisture deficit for early versus late snowmelt years, scaled by the fraction of forest cover in each grid cell, to map forests' vulnerability to changes in the timing of spring (Fig. 3) (26). The Northern Rockies and Northern California display the greatest vulnerability by this measure—the same forests accounting for more than three-quarters of increased wildfire frequency since the mid-1980s. Although the trend in temperature over the Northern Rockies increases with elevation, vulnerability in the Northern Rockies is highest around 2130 m, where the greatest increase in fires has occurred. At lower elevations, the moisture deficit in early years increased from a high average value (i.e., summer drought tends to be longer and more intense at lower elevations), whereas at higher elevations, the longer dry season in early years was still relatively short, and vegetation was somewhat buffered from the effects of higher temperatures by the available moisture.

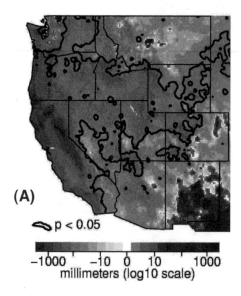

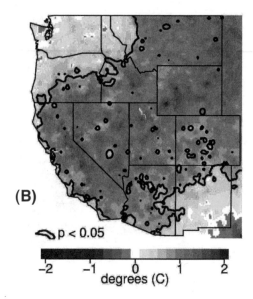

FIGURE 3. Average difference between early and late snowmelt years in average precipitation from October through May (A) and average temperature from March through August (B). Contours enclose regions in which a t test for the difference in mean between II early and II late years was significant (P < 0.05). The null hypothesis that precipitation from October through May is normally distributed could not be rejected using the Shapiro-Wilk test for normality (P > 0.05 for more than 95% of 24,170 grid cells, n = 49 for precipitation; P > 0.05 for more than 95% of 24,170 grid cells, n = 50 for temperature). See Fig. 1B for a definition of early, mid, and late snowmelt years.

Discussion

Robust statistical associations between wildfire and **hydroclimate** in western forests indicate that increased wildfire activity over recent decades reflects subregional responses to changes in climate. Historical wildfire observations exhibit an abrupt transition in the mid-1980s from a regime of infrequent large wildfires of short (average of one week) duration to one with much more frequent and longer-burning (five weeks) fires. This transition was marked by a shift toward unusually warm springs, longer summer dry seasons, drier vegetation (which provoked more and longer-burning large wildfires), and longer fire seasons. Reduced winter precipitation and an early spring snowmelt played a role in this shift. Increases in wildfire were particularly strong in mid-elevation forests.

The greatest absolute increase in large wildfires occurred in Northern Rockies forests. This subregion harbors a relatively large area of mesic, middle, and high-elevation forest types (such as lodgepole pine and spruce-fir) where fire exclusion has had little impact on natural fire regimes (1, 5) but where we found that an advance in spring produces a relatively large percentage increase in cumulative moisture deficit by

---

KEY TERM

Hydroclimate refers to aspects of the climate system (such as temperature and precipitation) that affect both the moisture available to plants to grow and the moisture available to wet the vegetation that fuels wildfires.

What's Already Happened?

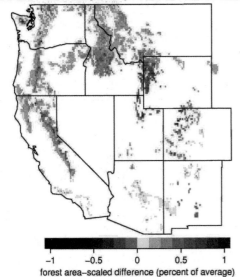

Forest Vulnerability: Early – Late Moisture Deficit

-1   -0.5   0   0.5   1
forest area–scaled difference (percent of average)

FIGURE 4. Index of forest vulnerability to changes in the timing of spring: the percentage difference in cumulative moisture deficit from October to September at each grid point in early versus late snowmelt years, scaled by the forest-type vegetation fraction at each grid point, for 1970 to 1999 (26). See fig. S3 online for a map of forest vulnerability for 1970 to 2003 over a smaller spatial domain. See Fig. 1B for a definition of early, mid, and late snowmelt years.

midsummer. In contrast, changes in Northern California forests may involve both climate and land-use effects. In these forests, large percentage changes in moisture deficits were strongly associated with advances in the timing of spring, and this area also includes substantial forested area where fire exclusion, timber harvesting, and succession after mining activities have led to increased forest densities and fire risks (10, 11). Northern California forests have had substantially increased wildfire activity, with most wildfires occurring in early years. Southwest forests, where fire exclusion has had the greatest effect on fire risks (2, 3), have also experienced increased numbers of large wildfires, but the relatively small forest area there limits the impact on the regional total, and the trend appears to be less affected by changes in the timing of spring. Most wildfires in the Southern Rockies and Southern California have also occurred in early snowmelt years, but again forest area there is small relative to the Northern Rockies and Northern California. Thus, although land-use history is an important factor for wildfire risks in specific forest types (such as some ponderosa pine and mixed-conifer forests), the broadscale increase in wildfire frequency across the western

United States has been driven primarily by sensitivity of fire regimes to recent changes in climate over a relatively large area.

The overall importance of climate in wildfire activity underscores the urgency of ecological restoration and fuels management to reduce wildfire hazards to human communities and to mitigate ecological impacts of climate change in forests that have undergone substantial alterations due to past land uses. At the same time, however, large increases in wildfire driven by increased temperatures and earlier spring snowmelts in forests where land-use history had little impact on fire risks indicates that ecological restoration and fuels management alone will not be sufficient to reverse current wildfire trends.

These results have important regional and global implications. Whether the changes observed in western hydroclimate and wildfire are the result of greenhouse gas–induced global warming or only an unusual natural fluctuation is beyond the scope of this work. Regardless of past trends, virtually all climate-model projections indicate that warmer springs and summers will occur over the region in coming decades. These trends will reinforce the tendency toward early spring snowmelt (36, 37) and longer fire seasons. This will accentuate conditions favorable to the occurrence of large wildfires, amplifying the vulnerability the region has experienced since the mid-1980s. The Intergovernmental Panel on Climate Change's consensus range of 1.5°C to 5.8°C projected global surface temperature warming by the end of the 21st century is considerably larger than the recent warming of less than 0.9°C observed in spring and summer during recent decades over the western region (37).

If the average length and intensity of summer drought increases in the Northern Rockies and mountains elsewhere in the western United States, an increased frequency of large wildfires will lead to changes in forest composition and reduced tree densities, thus affecting carbon ·pools. Current estimates indicate that western U.S. forests are responsible for 20% to 40% of total U.S. carbon sequestration (38, 39). If wildfire trends continue, at least initially, this biomass burning will result in carbon release, suggesting that the forests of the western United States may become a source of increased atmospheric carbon dioxide rather than a sink, even under a relatively modest temperature-increase scenario (38, 39). Moreover, a recent study has shown that warmer, longer growing seasons lead to reduced carbon dioxide uptake in high-elevation forests, particularly during droughts (40). Hence, the projected regional warming and consequent increase in wildfire activity in the western United States is likely to magnify the threats to human communities and ecosystems and substantially increase the management challenges in restoring forests and reducing greenhouse gas emissions (41).

References and Notes

1. C. Whitlock, *Nature* 432, 28 (2004).
2. W. W. Covington, *Nature* 408, 135 (2000).
3. C. D. Allen *et al.*, *Ecol. Appl.* 12, 1418 (2002).
4. J. L. Pierce, G. A. Meyer, A. J. T. Jull, *Nature* 432, 87 (2004).
5. T. Schoennagel, T. T. Veblen, W. H. Romme, *BioScience* 54, 661 (2004).
6. M. Savage, T. W. Swetnam, *Ecology* 71, 2374 (1990).
7. A. J. Belsky, D. M. Blumenthal, *Conserv. Biol.* 11, 315 (1997).
8. S. J. Pyne, P. L. Andrews, R. D. Laven, *Introduction to Wildland Fire* (Wiley, New York, 1996).
9. W. W. Covington, M. M. Moore, *J. For.* 92, 39 (1994).
10. K. S. McKelvey *et al.*, in *Sierra Nevada Ecosystems Project: Final Report to Congress*, X. X. Last, X. X. Last, Eds. (Univ. of California, Davis, CA, 1996), vol. 2, chap. 37.
11. G. E. Gruell, *Fire in Sierra Nevada Forests: A Photographic Interpretation of Ecological Change Since 1849* (Mountain Press, Missoula, MT, 2001).
12. T. Schoennagel, T. T. Veblen, W. H. Romme, J. S. Sibold, E. R. Cook, *Ecol. Appl.* 15, 2000 (2005).
13. R. C. Balling, G. A. Meyer, S. G. Wells, *Agric. For. Meteorol.* 60, 285 (1992).
14. E. K. Heyerdahl, L. B. Brubaker, J. K. Agee, *Holocene* 12, 597 (2002).
15. K. F. Kipfmueller, T. W. Swetnam, in *Wilderness Ecosystems, Threats, and Management*, D. N. Cole, S. F. McCool, W. T. Borrie, J. O'Loughlin, Eds. (U.S. Forest Service, RMRS-P-15, Fort Collins, CO, 2000), vol. 5, pp. 270–275.
16. T. W. Swetnam, J. L. Betancourt, *J. Clim.* 11, 3128 (1998).
17. T. T. Veblen, T. Kitzberger, J. Donnegan, *Ecol. Appl.* 10, 1178 (2000).
18. A. L. Westerling, T. J. Brown, A. Gershunov, D. R. Cayan, M. D. Dettinger, *Bull. Am. Meteorol. Soc.* 84, 595 (2003).
19. T. W. Swetnam, J. L. Betancourt, *Science* 249, 1017 (1990).
20. A. Gershunov, T. P. Barnett, *J. Clim.* 11, 1575 (1998).
21. A. Gershunov, T. P. Barnett, D. R. Cayan, *Eos* 80, 25 (1999).
22. A. L. Westerling, T. W. Swetnam, *Eos* 84, 545 (2003).
23. N. L. Stephenson, *Am. Nat.* 135, 649 (1990).
24. N. L. Stephenson, *J. Biogeogr.* 25, 855 (1998).
25. T. W. Swetnam, *Science* 262, 885 (1993).
26. Materials and methods are available as supporting material on *Science* Online.
27. K. E. Mitchell *et al.*, *J. Geophys. Res.* 109, D07S90 (2004).
28. E. P. Maurer, A. W. Wood, J. C. Adam, D. P. Lettenmaier, B. Nijssen, *J. Clim.* 15, 3237 (2002).
29. A. F. Hamlet, D. P. Lettenmaier, *J. Hydrometeor.* 6, 330 (2005).
30. NCDC, *Time Bias Corrected Divisional Temperature-Precipitation-Drought Index*, documentation for data set TD-9640 (DBMB, NCDC, NOAA, Asheville, NC, 1994); www.ncdc.noaa.gov/oa/climate/onlineprod/drought/readme.html.

31. X. Liang, D. P. Lettenmaier, E. F. Wood, S. J. Burges, *J. Geophys. Res.* 99, 14415 (1994).

32. I. T. Stewart, D. R. Cayan, M. D. Dettinger, *J. Clim.* 18, 1136 (2005).

33. D. R. Cayan, S. A. Kammerdiener, M. D. Dettinger, J. M. Caprio, D. H. Peterson, *Bull. Am. Meteorol. Soc.* 82, 399 (2001).

34. J. R. Slack, J. M. Landwehr, *U.S. Geol. Surv. Open-File Rep. 92–129* (1992).

35. J. Sheffield, G. Goteti, F. H. Wen, E. F. Wood, *J. Geophys. Res.* 109, D24108 (2004).

36. National Assessment Synthesis Team, *Climate Change Impacts on the United States: The Potential Consequences of Climate Variability and Change* (U.S. Global Change Research Program, Washington, DC, 2000).

37. J. T. Houghton *et al.*, Eds., *IPCC Climate Change 2001: The Scientific Basis* (Cambridge Univ. Press, Cambridge, 2001).

38. S. W. Pacala *et al.*, *Science* 292, 2316 (2001).

39. D. Schimel, B. H. Braswell, in *Global Change and Mountain Regions: An Overview of Current Knowledge*, U. M. Huber, H. K. M. Bugmann, M. A. Reasoner, Eds., vol. 23 of *Advances in Global Change Research* (Springer, Dordrecht, Netherlands, 2005), pp. 449–456.

40. W. Sacks, D. Schimel, R. Monson, *Oecologia*, in press.

41. We thank M. Dettinger and D. Schimel for help. This work was supported by grants from the National Oceanographic and Atmospheric Administration's Office of Global Programs, the National Fire Plan by means of the U.S. Forest Service's Southern Research Station, and the California Energy Commission.

Supporting Online Material

www.sciencemag.org/cgi/content/full/1128834/DC1

Materials and Methods
Figs. S1 to S3
References

How Much More Rain Will Global Warming Bring?

FRANK J. WENTZ, LUCREZIA RICCIARDULLI,
KYLE HILBURN, CARL MEARS

Climate models and satellite observations both indicate that the total amount of water in the atmosphere will increase at a rate of 7% per kelvin (K) of surface warming. However, the climate models predict that global precipitation will increase at a much slower rate of 1% to 3% per K. A recent analysis of satellite observations does not support this prediction of a muted response of precipitation to global warming. Rather, the observations suggest that precipitation and total atmospheric water have increased at about the same rate over the past two decades.

In addition to warming Earth's surface and lower troposphere, the increase in greenhouse gases (GHG) concentrations is likely to alter the

This article first appeared in *Science* (Originally published in *Science* Express on 31 May 2007, *Science* 13 July 2007: Vol. 317, no. 5835). It has been revised for this edition.

> ## KEY TERM
>
> The troposphere is the lowest atmospheric layer, closest to Earth's surface. It ranges from 4 to 11 miles high, depending on the latitude at which it is measured.

planet's hydrologic cycle (1–3). If the changes in the intensity and spatial distribution of rainfall are substantial, they may pose one of the most serious risks associated with climate change. The response of the hydrologic cycle to global warming will depend to a large degree on the way in which the enhanced GHG will alter the radiation balance in the troposphere. As GHG concentrations increase, the climate models predict, an enhanced radiative cooling will be balanced by an increase in latent heat from precipitation (1, 2). The Coupled Model

Intercomparison Project (4) and similar modeling analyses (1–3) predict a relatively small increase in precipitation (and likewise in evaporation) at a rate of about 1% to 3% per K of surface warming. In contrast, both climate models and observations indicate that the total water vapor in the atmosphere increases by about 7% per K (1–3, 5, 6).

More than 99% of the total moisture in the atmosphere is in the form of water vapor. The large increase in water is due to the ability of the warmer air to hold more water vapor, as dictated by the **Clausius-Clapeyron (C-C) relation** under the condition that the relative humidity in the lower troposphere stays constant. So according to the current set of global coupled ocean-atmosphere models (GCMs), the rate of increase in precipitation will be several times lower than that for total water. This apparent inconsistency is resolved in the models by a reduction in the vapor mass flux (total mass of water vapor flowing through a surface), particularly with respect to the Walker circulation, which reinforces the trade winds (3, 7). Whether a decrease in global winds is a necessary consequence of global warming is a complex question that is yet to be resolved (8).

Using satellite observations from the Special Sensor Microwave Imager (SSM/I), we assessed the GCMs' prediction of a muted response of rainfall and evaporation to global warming. The SSM/I is well suited for studying the global hydrologic cycle in that it simultaneously measures precipitation (P), total water vapor (V), and also surface-wind stress (τ_0, the horizontal force of the wind on the ocean surface), which is the principal term in the computation of evaporation (E) (8, 9).

The SSM/I data set extends from 1987 to 2006. During this time, Earth's surface temperature warmed by 0.19 ± 0.04 K per decade, according to the Global Historical Climatology Network (10, 11). Satellite measurements of the lower troposphere show a similar warming of 0.20 ± 0.10 K per decade (12). The error bars are at the 95% confidence level. This warming is consistent with 20th-century climate-model runs (13) and provides a reasonable, albeit short, test bed for assessing the impact of global warming on the hydrologic cycle.

When averaged globally over monthly time scales, P and E must balance except for a negligibly small storage term. This $E = P$ constraint provides a useful consistency check with which to evaluate our results. However, the constraint is valid only for global averages. Accordingly, the first step in our analysis was to construct global monthly maps of P and E at a 2.5° spatial resolution for the period 1987 to 2006.

The SSM/I retrievals used here are available only over the ocean. To supplement the SSM/I over-ocean rain retrievals, we used the land values from the Global Precipitation Climatology Project data set, which is a blend of satellite retrievals and rain gauge measurements (14, 15). Satellite rain retrievals over land were less accurate than their ocean counterparts, but this drawback was compensated for by the fact that there are abundant rain gauges over land for constraining the satellite retrievals. Likewise, global evaporation was computed separately for oceans and land. Because 86% of the world's evaporation comes from the oceans (16), ocean evaporation was our primary focus. We computed evaporation over the oceans with the use of the bulk formula from the National Center for Atmospheric Research Community Atmospheric Model 3.0 (8, 17). Evaporation over land cannot be derived from satellite observations, and we resorted to using a constant value of 527 mm/year for all of the continents, excluding Antarctica (16). For Antarctica and sea ice, we used a value of 28 mm/year (8, 16).

The GCMs indicate that E should increase by about 1% to 3% per K of surface warming. However, according to the bulk formula (see eq. S1 online) (8), evaporation increases similarly to C-C as the surface temperature warms, assuming that the other terms remain constant. For example, a global increase of 1 K in the surface air temperature produces a 5.7% increase in E (8). For a muted response of 1% to 3% per K, other variables in the bulk formula need to change with time. The air-sea temperature difference and the near-surface relative humidity are expected to remain nearly constant (8), and this leaves τ_0 as the one variable that can reduce evaporation to the magnitude required to balance the radiation budget in the models. To bring the bulk formula into agreement with the radiative cooling constraint, would need to decrease by about 4% per K. Thus, a muted response of precipitation to global warming requires a decrease in global winds (2, 3, 7).

To evaluate the GCMs' prediction of a decrease in winds, we looked at the 19 years of SSM/I wind retrievals. These winds are expressed in terms of an equivalent neutral-stability wind speed (W) at a 10 m elevation, which is proportional to (8, 16). Figure 1 shows a decadal trend map of W. For each 2.5° grid cell, a least-squares linear fit to the 19-year time series was calculated after removing the seasonal variability. The wind trends from the International Comprehensive Ocean-Atmosphere Data Set (ICOADS) are also shown, but just for comparison. They were not used in our analysis. Although the ICOADS trend map is very noisy because of sampling and measurement deficiencies, it shows trends similar to those from the SSM/I in the northern Atlantic and Pacific, where the ICOADS ship observations are more abundant. The North Atlantic Oscillation (NAO) is apparent in both trend maps as a tripole feature with increasing winds between 30°N and 40°N and decreasing winds to the north and south (18). This feature is consistent with the observed decrease in the NAO index since 1987. When averaged over the tropics from 30°S to 30°N, the winds increased by 0.04 m/s (0.6%) per decade, and over all oceans, the increase was 0.08 m/s (1.0%) per decade. The SSM/I

SSM/I Wind Trend Map

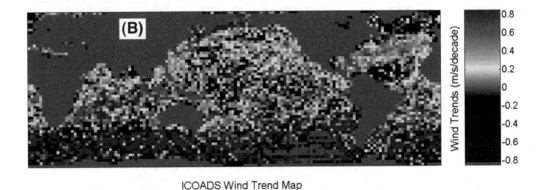

ICOADS Wind Trend Map

FIGURE 1. Surface-wind trends for the period July 1987 through August 2006 computed at a spatial resolution of 2.5°. (A) SSM/I wind trends. (B) ICOADS wind trends. In the North Pacific and North Atlantic where ICOADS ship observations are more abundant, the two data sets show similar trends. The tripole feature in the North Atlantic is consistent with the recent decrease in the NAO index.

wind retrievals were validated by comparisons with moored ocean buoys and satellite scatterometer wind retrievals (a way of measuring surface-wind stress, fig. S1). On the basis of this analysis, the error bar on the SSM/I wind trend is estimated to be ± 0.05 m/s per decade at the 95% confidence level (8). This observed increase in the SSM/I winds is opposite to the GCM results, which predict that the 1987–2006 warming should have been accompanied by a decrease in winds on the order of (0.19 K per decade)(4% per K) = 0.8% per decade.

We then looked at the variability of global precipitation and evaporation during the past two decades. Figure 2A shows the time series for P and E. Also shown is the over-ocean SSM/I

retrieval of V. For the time series generation, the seasonal variability was first removed, and then the variables were low-pass filtered (so that high frequencies were filtered out) by convolution with a Gaussian distribution (a symmetric bell-shaped probability distribution) that had a ± four-month width at half-peak power. The major features apparent in the time series are the 1997–1998 El Niño and the 1986–1987 El Niño, followed by the strong 1988–1989 La Niña. It is noteworthy that E, P, and V all exhibited similar magnitudes for interannual variability and decadal trends (Table 1). After applying the ± four-month smoothing, the correlation of E versus P was 0.68. Because global precipitation and evaporation must balance, the

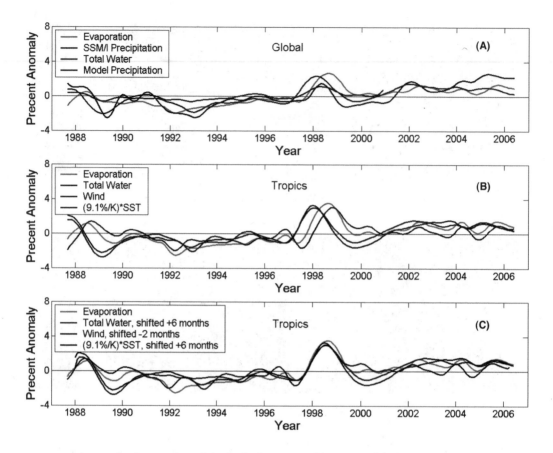

FIGURE 2. Anomaly time series of the hydrologic variables. (A) Global results for the observed precipitation and evaporation and over-ocean results for total water vapor. The average model precipitation predicted by AMIP simulations is also shown. (B) Tropical ocean results for evaporation, total water vapor, surface wind speed, and SST. The SST time series has been scaled by 9.1% per K. During the El Niños, evaporation and wind were not in phase with vapor and SST. At the end of 1996, SST and vapor began to increase while the winds began to decrease, with no net effect on evaporation. About eight months later (mid-1997), the winds in the tropics began to recover and then increased sharply, reaching a maximum value in late 1998. All four variables remained at elevated values thereafter. (C) Same as (B), except that the water vapor and SST curves have been shifted forward in time by six months, and the wind curve has been shifted backward by two months. The statistics on the global time series, including error bars, are given in Table 1.

observed differences in the derived values of P and E provided an error estimate that we used to estimate the uncertainty in the decadal trend. The estimated error bar at the 95% confidence level for E and P is ± 0.5% per decade (8).

Also shown in Figure 2A is the ensemble mean of nine climate-model simulations smoothed in the same way as in the satellite observations. These climate runs, for which

the sea surface temperature (SST) is prescribed from observations, are from the Atmospheric Model Intercomparison Project II (AMIP-II) (19, 20). There is a pronounced difference between the precipitation time series from the climate models and that from the satellite observations. The amplitude of the interannual variability, the response to the El Niños, and the decadal trends are all smaller by a factor of two to three

TABLE 1. Statistics on the variation of global evaporation, global precipitation, and over-ocean water vapor for the period July 1987 through August 2006. The error bars on the trends are given at the 95% confidence level. The values in parentheses are in terms of percentage change, rather than absolute change.

Parameter	Mean	Standard deviation	Trend
Evaporation	961 mm year^{-1}	10.1 mm year^{-1} (1.1%)	12.6 ± 4.8 mm year^{-1} decade^{-1} (1.3 ± 0.5% decade^{-1})
Precipitation	950 mm year^{-1}	12.7 mm year^{-1} (1.3%)	13.2 ± 4.8 mm year^{-1} decade^{-1} (1.4 ± 0.5% decade^{-1})
Total water	28.5 mm	0.292 mm (1.0%)	0.354 ± 0.114 mm decade^{-1} (1.2 ± 0.4% decade^{-1})

in the climate-model results, as compared with the observations. This characteristic of the models to underpredict the amplitude of precipitation changes to El Niño–Southern Oscillation events has been reported previously (21), and the results presented here suggest that the climate models are also underestimating the decadal variability.

The similarity in the satellite-derived time series became more pronounced when the analysis was limited to the tropical oceans (30°S to 30°N), where most of the evaporation occurs. Although the condition $E = P$ was no longer valid for this regional analysis, the coupling of evaporation with the other variables was more apparent. Figure 2B shows the tropical time series of E, V, SST, and W. The variables V and SST exhibited a high correlation [correlation coefficient (r) = 0.96], and their scaling relation of 9.1% per K was equal to the C-C rate (6.5% per K) times a moist adiabatic lapse rate [(MALR), the rate of decrease of temperature with height for air saturated with water vapor at 100% rela-

tive humidity] factor of 1.4 (5). The MALR factor is the ratio of change in the lower tropospheric temperature to the change in SST. This strong coupling between V and SST is another confirmation that the total atmospheric water increases with temperature at the C-C rate.

During the two El Niños, evaporation and wind speed were not in phase with vapor and SST. The increase in evaporation lagged the increase in vapor by six months, and the increase in winds lagged by eight months (Fig. 2B). When a six-month lag was applied, the correlation between E and V was 0.84 in the tropics and 0.88 globally.

Figure 3 shows a trend map of $P - E$. The most striking feature is in the tropical Western Pacific Warm Pool, where $\Delta (P - E)$ is about 400 mm/year per decade, and Δ represents change. This is a region of maximum $P - E$ (1500 to 2000 mm/year). Simple hydrologic models predict that $\Delta (P - E)$ should vary similarly to $P - E$ (3). That is to say, wet regions should get wetter and dry regions should get drier. This seems to be

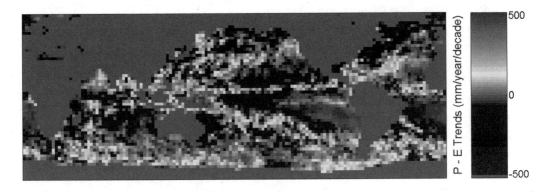

FIGURE 3. Trends in satellite-derived $P - E$ for the period July 1987 through August 2006. The largest change was over the Warm Pool in the Western Pacific—a wet area that became wetter.

the case over the Warm Pool, but elsewhere this direct proportionality is not as apparent.

During the past two decades, the hydrologic parameters E, P, and V exhibited similar responses to the two El Niños (apart from a six-month lag), similar magnitudes of interannual variability (1.0% to 1.3%), and similar decadal trends (1.2% to 1.4% per decade). Earth's surface warmed by 0.2 K per decade during this period, and hence the observed changes in E and P suggest an acceleration in the hydrologic cycle of about 6% per K, close to the C-C value. In addition, ocean winds exhibited a small increase of 1.0% per decade. There is no evidence in the observations that radiative forcing in the troposphere is inhibiting the variations in E, P, and W. Rather, E and P seem to simply vary in unison with the total atmospheric water content.

The reason for the discrepancy between the observational data and the GCMs is not clear. One possible explanation is that two decades is too short a time period, and thus we see internal climate variability that masks the limiting effects of radiative forcing. However, we would argue that although two decades may be too short for extrapolating trends, it may indeed be long enough to indicate that the observed scaling relations will continue on a longer time scale. Another possible explanation is that there are errors in the satellite retrievals, but the consistency among the independent retrievals and validation of the winds with other data sets

suggests otherwise. Finally, there is the possibility that the climate models have in common a compensating error in characterizing the radiative balance for the troposphere and Earth's surface. For example, variations in modeling cloud-radiative forcing at the surface can have a relatively large effect on the precipitation response (4), whereas the temperature response is more driven by how clouds affect the radiation at the top of the troposphere.

The difference between a subdued increase in rainfall and a C-C increase has enormous impact, with respect to the consequences of global warming. Can the total water in the atmosphere increase by 15% with carbon dioxide doubling but precipitation increase by only 4% (1)? Will warming really bring a decrease in global winds? The observations reported here suggest otherwise, but clearly these questions are far from being settled (22).

References and Notes

1. M. R. Allen, W. J. Ingram, *Nature* 419, 224 (2002); http://www.sciencemag.org/cgi/external_ref?access_num=12226677&link_type=MED.
2. J. F. B. Mitchell, C. A. Wilson, W. M. Cunnington, *Q. J. R. Meteorol. Soc.* 113, 293 (1987).
3. I. M. Held, B. J. Soden, *J. Clim.* 19, 5686 (2006).
4. C. Covey *et al.*, *Global Planet. Change* 37, 103 (2003).
5. F. J. Wentz, M. Schabel, *Nature* 403, 414 (2000).
6. K. E. Trenberth, J. Fasullo, L. Smith, *Clim. Dyn.* 24, 741 (2005).
7. G. A. Vecchi *et al.*, *Nature* 441, 73 (2006).
8. See supporting material on *Science* Online.
9. F. J. Wentz, *J. Geophys. Res.* 102, 8703 (1997).
10. T. M. Smith, R. W. Reynolds, *J. Clim.* 18, 2021 (2005).
11. Data are posted at www.ncdc.noaa.gov/oa/climate/research/ghcn/ghcngrid.html.
12. C. A. Mears, F. J. Wentz, *Science* 309, 1548 (2005); published online 11 August 2005 (10.1126/science.1114772).
13. G. A. Meehl *et al.*, *Science* 307, 1769 (2005).
14. R. F. Adler *et al.*, *J. Hydrometeorol.* 4, 1147 (2003).
15. The satellite-gauge combined precipitation product V2 was downloaded from ftp://precip.gsfc.nasa.gov/pub/gpcp-v2/psg/.
16. J. P. Peixoto, A. H. Oort, *Physics of Climate* (Springer, New York, 1992), pp. 170–172, 228–237.

17. W. D. Collins *et al., NCAR Technical Note TN-464+STR* (National Center for Atmospheric Research, Boulder, CO, 2004).

18. J. W. Hurrell, Y. Kushnir, G. Ottersen, M. Visbeck, Eds., *The North Atlantic Oscillation: Climate Significance and Environmental Impact,* Geophysical Monograph Series (American Geophysical nion, Washington, DC, 2003), pp. 1–35.

19. W. L. Gates *et al., Bull. Am. Meteorol. Soc.* 73, 1962 (1998).

20. The more recent AMIP-II simulations were obtained from www-pcmdi.llnl.gov/projects/amip.

21. B. J. Soden, *J. Clim.* 13, 538 (2000).

22. This work was supported by NASA's Earth Science Division.

Supporting Online Material

www.sciencemag.org/cgi/content/full/1140746/DC1
Materials and Methods
SOM Text
Fig. S1
References

Running Out of Water— and Time

Geography, politics, and war combine to make the Gaza Strip a worst-case scenario for water-resource planners.

JOHN BOHANNON

RAFAH—You can almost hear the collective sigh of relief as the angry Sun sets over this dusty city on Gaza's Egyptian border. This is when five of Ali Abu Taha's sons arrive, unwinding their kaffiyehs and gathering around the charcoal fire where a pot of tea is already boiling. The unprecedented visit of a foreign guest calls for a demonstration of the hospitality for which the Bedouins are famous. The seat of honor is offered, and some of the family's most valuable possessions are laid out on the carpet for display: a battered old AK-47 rifle and several bottles of home-filtered water. The gun is a family heirloom that rarely sees light, but the water is indispensable. "The filter cartridges are very expensive and hard to get into Gaza," says one of the sons,

Mohammed, "but this one should hold up for another month, insha' Allah." Not only does it provide the drinking water for Abu Taha's clan—about 100 people, a third of them his grandchildren—but by enabling them to bottle and sell water to neighbors, it provides one of their few sources of income.

"I don't recommend drinking too much of this," says Mohammed as he fills a glass with unfiltered water from the tap. One sip of the pongy (smelly) brine is enough to understand why. As a general rule, the farther south one goes in Gaza, the worse the water becomes, and Rafah is the end of the line. The Palestinian Authority issues warnings from time to time urging the public to buy bottled water, especially for the very young or elderly. But for the average Gazan—with an annual income of $600—a $1 gallon of water is a luxury.

For Abu Taha, who grew up as a Bedouin in the nearby Negev desert, making efficient use of scarce water resources is nothing new. The

This article first appeared in *Science* (25 August 2006: Vol. 313, no. 5790). It has been revised for this edition.

problem is that the 1.4 million people crammed into the Gaza Strip—most of them the children of refugees who fled their homes in the 1948 and 1967 Arab-Israeli wars—depend on a shallow aquifer for water. But year by year, that source is becoming more contaminated by salt and pollution. Most wells already produce water that is nonpotable by standards set by the (WHO) Organization.

Water scarcity is a perennial problem in the region, but nowhere is it worse than in Gaza. "It is a microcosm of the entire Middle East," says Eric Pallant, an environmental scientist at Allegheny College in Meadville, Pennsylvania, who has collaborated with both Israelis and Palestinians on water problems. "If you can figure out how to make water sustainable there, then you can do it anywhere." Several Gaza water projects have been planned by donor countries in recent years, including state-of-the-art wastewater treatment and desalination plants, but all have fizzled because of security concerns and sanctions slapped onto the new Hamas-led Palestinian government. Israel's withdrawal of settlers and troops from Gaza last year is a bittersweet victory for the Palestinians. Although they are fully in control of Gaza's water for the first time, they must now scramble to save it before it becomes irreversibly contaminated.

Water Woes

It is a tense first day on the job for Mohammad Al-Agha, the Hamas minister of agriculture. Like a thunderstorm that never quite arrives, Israeli artillery pounds the landscape to the north where Palestinian militants have been launching rockets over the border. After a brief welcome party in his new Gaza City office, Al-Agha, a geologist from nearby Islamic University, and the small group of experts responsible for managing Gaza's water resources meet with *Science* to discuss their plans. The conversation is interrupted twice when fighter jets scream overhead and strike nearby targets with missiles, causing the building to shudder.

IMAGE 1: Drink at your own risk. With salinity and pollution on the rise, a slim and shrinking minority of Gaza's wells meet health standards.

At a glance, the Gazans' water woes seem insurmountable. The only natural fresh source available is the coastal aquifer, a soggy sponge of sediment layers that slopes down to the sea a few dozen meters beneath their feet (see figure online). Its most important input is the meager 20 to 40 cm of annual rainfall that sprinkles over Gaza's 360 km^2 surface—about twice the area of Washington, D.C.—giving between 70 and 140 million cubic meters (MCM) of water per year. Most of that water evaporates, but between 20 and 40 MCM penetrates the sandy sediment to feed the aquifer. Another 15 to 35 MCM, depending on whom you ask, flows in under the border from Israel, while irrigation and leaky pipes are estimated to return 40 to 50 MCM, for a total annual recharge of 75 to 125 MCM.

The aquifer's only natural output is the 8 MCM per year that should exit into the Mediterranean, providing a crucial barrier against

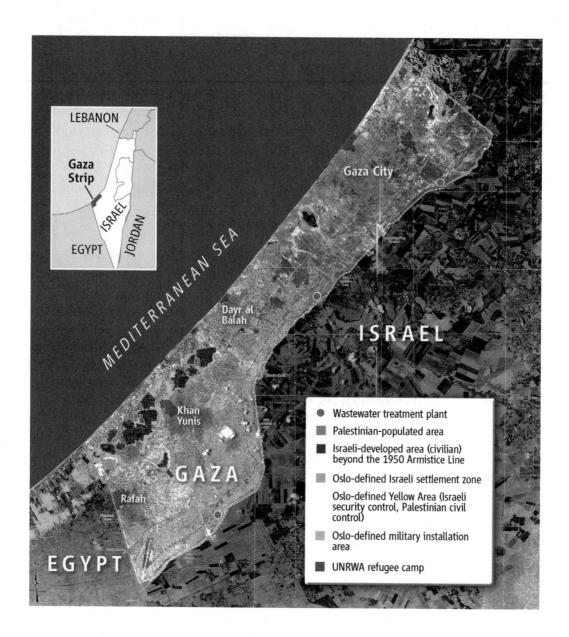

IMAGE 2: Urban pollution, agricultural contamination, and saltwater intrusion from the Mediterranean Sea give Gaza's aquifer a dicey future. Sources: University of Texas; UNEP-GRID, Geneva. (http://www.sciencemag.org/cgi/content/full/313/5790/1085/F2.)

the intrusion of seawater. So if no more than about 100 MCM were tapped from the aquifer per year, it could last forever. But Gaza's 4000 wells suck out as much as 160 MCM yearly, says Ahmad Al-Yaqoubi, a hydrologist who directs the Palestinian Water Authority. This estimated 60 MCM annual water deficit is why the water table is dropping rapidly and already reaches 13 meters below sea level in some places. Salt water from the Mediterranean as well as deeper pockets of brine get sucked in to fill the gap. "The saltwater intrusion is well under way," says Al-Yaqoubi, "especially in the coastal areas and to the south." About 90% of wells already have salinity exceeding the WHO-recommended maximum of 250 parts per million (ppm). The accelerating rate of saltwater intrusion alone could make the Gaza aquifer unusable within two or three decades, according to a 2003 report by the United Nations Environment Programme.

But there may be far less time on the clock. The aquifer is also mixing with a cocktail of pollutants from Gaza's sewage and agriculture. "Besides salt, our number-one contaminant is nitrate from solid waste and fertilizers," says Yousef Abu Safieh, an environmental scientist based in Gaza City who heads the Palestinian Environmental Quality Authority. The maximum safe concentration of nitrate according to WHO is 45 ppm. "In our sampling, we find that most wells have about 200 ppm, and wells close to agricultural runoff can even hit 400," says Abu Safieh. Two Palestinian governmental studies led by Abu Safieh point to patterns of disease matching the distribution of water contamination. The higher the salinity of local water, the higher the incidence of kidney disease, he says, and nitrate concentration correlates with Gaza's high incidence of blue baby syndrome—a loss of available oxygen in the blood that can cause mental retardation or be fatal.

It is the job of a water utility to clean up such contamination and make sure that safe water comes out of the tap, but there is no such unified utility in Gaza. Instead, the Strip is covered by a patchwork of fragmented water infrastructure. Gaza's three wastewater treatment plants are far from adequate. The largest, south of Gaza City, was designed to treat 42,000 m^3 per day—the amount produced by 300,000 people—but now faces a daily inflow of more than 60,000 m^3, says Al-Yaqoubi: "This has overwhelmed the biological step of the treatment process." As an emergency measure to prevent sewage from overflowing, barely treated wastewater is now piped to the coast, where the dark gray liquid can be seen, and smelled, flowing along the beach. Meanwhile, the 40% of Gazans without access to a centralized sewage disposal system contribute to the burgeoning cesspits. A 40 ha lake of sewage that has formed in northern Gaza

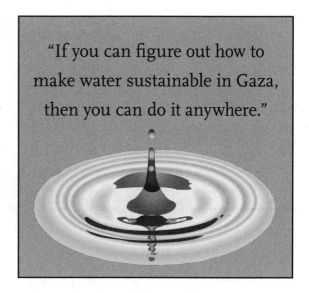

"If you can figure out how to make water sustainable in Gaza, then you can do it anywhere."

is a menace to people at the surface and the aquifer beneath.

These threats to the water supply are serious, says Al-Yaqoubi, but "water scarcity is of course the problem that will never go away." Considering that crop irrigation gobbles up 70% of Gaza's water and fertilizers contribute most of the nitrate contamination, firmer control of agriculture by Al-Agha's ministry seems like a necessary first step in saving the aquifer from ruin. "The problems continue to spiral," says Mac McKee, a hydrologist at Utah State University in Logan, who has collaborated with Gazans for the past 10 years, because "the Palestinian Authority has not succeeded in applying effective controls on well drilling and pumping." About half of Gaza's wells have been dug illegally, mostly by farmers to irrigate small plots of cropland.

"If you try to tell farmers to stop using their wells, they come out with guns," says Ehab Ashour, a water engineer who works for international development agencies in Gaza. And with the struggle for power intensifying between the Hamas and Fatah leaderships, the prospect of better enforcement seems dimmer than ever. For his part, Al-Agha says a crackdown on well digging isn't even on the table. "We can't do this from an economic standpoint," he says. "Over 60% of people here are farming. We are all

IMAGE 3: Sewage gushes out onto Gaza's beaches from a failing wastewater treatment plant. Plans to upgrade the infrastructure remain frozen.

locked into this jail, so we have to grow our own food and at the same time try to produce something we can export."

Asking families in Gaza to use less water is also "out of the question," says David Brooks, an environmental scientist who was an adviser during the Israeli-Palestinian water negotiations—a mandate of the Oslo Peace Accords—until negotiations collapsed in 2003 and who is now at Friends of the Earth in Ottawa, Canada. Average daily domestic water consumption in Gaza is about 70 liters per person—used not only in homes but also hospitals, schools, businesses, and public institutions—whereas 100 liters per capita per day is the generally agreed minimum for public health and hygiene. (By comparison, average consumption in Israel is 280 liters per day.) So the only way forward is to secure new sources of fresh water and make existing

sources stretch further, says Abu Safieh. There are several strategies for doing this, he says, "and we are pursuing all of them."

Making More with Less

If you stand on any hill in Gaza and look west, a tantalizing source of water shimmers into view. If only the salts could be efficiently removed, the Mediterranean is a virtually limitless supply for desalination plants. Indeed, this very water is feeding some of the world's most advanced facilities less than an hour's drive up the coast in Israel.

For any long-term solution in Gaza, "desalination will be absolutely necessary," says McKee. A desalination plant capable of providing Gaza with 60 MCM of drinking water per year was part of a plan drawn up by the U.S. Agency for

International Development (USAID) in 2000. Money to build the $70 million plant, along with $60 million to lay down a carrier system to pipe the water across Gaza, was ready to go from USAID when the second intifada broke out just months later, stalling the project. It was officially frozen in 2003 after a bombing killed three members of a U.S. diplomatic convoy in Gaza.

Besides producing more drinking water, the priority is to deal with Gaza's sewage, says Al-Yaqoubi, not only to prevent a public health disaster, but also to recycle some of the precious water back into the system. A trio of wastewater treatment plants that could handle Gaza's entire load has been promised by USAID, the World Bank, Germany, Finland, and Japan, but "nothing has happened," he says, because of the Hamas election victory.

In relation to stretching the current water supply further, there is one positive legacy of Israeli occupation in Gaza. By working on Israeli farms, "we have become very comfortable with new technologies," says Al-Yaqoubi. In spite of the official freeze on international aid to the Palestinian government, projects aiming to improve farming in Gaza "are ongoing by many donors," he says. The most important is drip irrigation, delivering water directly to roots through a network of tubes. Coupling this with a computerized system that automatically pumps just enough water from a well to meet the plants' daily needs can make irrigation up to 70% more efficient over the long run.

But for the immediate crisis, the country best placed to help Gaza may be Israel. Before the taps were shut this year after Hamas was elected, 5 MCM per year of drinking water was being piped into Gaza by Mekorot, the Israeli national water company, and an additional 5 MCM had been agreed on. That water does not come free, but it is nevertheless a freshwater source separate from the ailing aquifer.

"We know how serious the situation is in Gaza," says Saul Arlosoroff, a member of Mekorot's board of directors and a former Israeli deputy water commissioner. "The first priority

> The first priority is to get these people enough clean drinking water, and the second is to prevent salinity from irreversibly destroying their soil."

is to get these people enough clean drinking water, and the second is to prevent salinity from irreversibly destroying their soil." Arlosoroff says Israelis and Palestinians working in the water sector have a special relationship. "We understand each other, and we know that these problems require cooperation," he says, "but the atmosphere between Gaza and Israel is worse now than at any time in our history."

Across the border, Abu Safieh is similarly disappointed. "There was a time when I could talk with my Israeli counterpart constructively about our environmental problems," he says, but he has not had any contact in years. Al-Agha says he plans to turn to Egypt for help. For importing and exporting, as well as perhaps for obtaining the abundant electricity needed to desalinate water, he says, "our hope is to the south."

The present turmoil also prevents what Brooks calls "the easiest and best solution" to Gaza's environmental problems: reducing the number of people living there. "Gaza can't sustain that population, and any real solution will require people to leave," he says. Most Gazans "will never give up hope of returning to their homes," says Abu Safieh, but for now, "we will work to make the best of the bad situation."

Rapid Advance of Spring Arrival Dates in Long-Distance Migratory Birds

NICLAS JONZÉN, ANDREAS LINDÉN, TORBJØRN ERGON,
ENDRE KNUDSEN, JON OLAV VIK, DIEGO RUBOLINI,
DARIO PIACENTINI, CHRISTIAN BRINCH, FERNANDO SPINA,
LENNART KARLSSON, MARTIN STERVANDER, ARNE ANDERSSON,
JONAS WALDENSTRÖM, ALEKSI LEHIKOINEN, ERIK EDVARDSEN,
RUNE SOLVANG, NILS CHR. STENSETH

Several bird species have advanced the timing of their spring migration in response to recent climate change. European short-distance migrants, wintering in temperate areas, have been assumed to be more affected by change in the European climate than long-distance migrants wintering in the tropics. However, we show that long-distance migrants have advanced their spring arrival in Scandinavia more than short-distance migrants. By analyzing a long-term data set from southern Italy, we show that

This article first appeared in *Science* (30 June 2006: Vol. 312, no. 5782). It has been revised for this edition.

long-distance migrants also pass through the Mediterranean region earlier. We argue that this may reflect a climate-driven evolutionary change in the timing of spring migration.

Many biological processes are affected by climate, and in temperate areas, the increasing spring temperature over the past 20 to 30 years has caused an advancement of phenological events in plants and invertebrates (1, 2). The earlier onset of spring has consequences for the timing of breeding in birds, which has evolved to match peak food availability (3, 4). We may therefore expect the timing of breeding to track any temporal shift in food availability caused by a trend in spring temperature (5). Most passerine birds (which constitute more than half of all bird species) breeding in temperate areas of the Northern Hemisphere are seasonal migrants, and the timing of migration ultimately constrains when breeding can start (6, 7). Short-distance migrants, spending the winter close to the breeding grounds, may be able to adjust the timing of migration in response to local climate change, which will be correlated to the conditions on the breeding grounds. In tropical-wintering long-distance migrants, the timing of migration is under endogenous control (8, 9), and the cues needed to trigger the onset of migration are unlikely to be linked to the climate on their breeding grounds. Therefore, it has been assumed that short-distance migrants are more likely than long-distance migrants to vary migration timing in response to climate change (10). Here we show that such an assumption is not empirically justified.

We estimated trends in arrival time for the early, middle, and late phases of migration (that is, the species- and site-specific 10th, 50th, and 90th percentiles of the spring arrival distribution) in short- and long-distance passerine migrants, based on long-term banding and observational data (from 1980 to 2004) from four bird observatories in Scandinavia and a site in southern Italy (11). We also investigated whether year-to-year variation in arrival time can be explained by short-term climate variability as measured by the North Atlantic Oscillation (NAO) (12). As explanatory variables, we used the calendar year (*TIME*) and the deviations from linear regression of the winter NAO index on year (*dNAO*) [the trend in NAO was weakly negative over this time period (11)]. Spring

migration might advance for two distinct reasons. First, there can be a microevolutionary (genetic) response to the selection pressures for earlier breeding. Second, the migrants can show a phenotypically plastic response to trends in weather or climatic patterns on the wintering ground and/or along the migration route, whereby if spring arrives early on the wintering grounds, spring migration will also start early. Thus, a response to *TIME* may reflect either microevolutionary change or phenotypic plasticity, whereas a response to *dNAO* indicates exclusively phenotypic plasticity in the migratory behavior.

Long-distance migrants have advanced their arrival in northern Europe in all phases of migration (see tables S1 to S3 online and Fig. 1). The advancement in long-distance migrants is strongest in the early phase of migration, and there is limited variation between species. Furthermore, the analysis of the data set from Italy (from the island of Capri) showed that long-distance migrants wintering south of the Sahara are actually arriving in southern Europe progressively earlier. In fact, all of the nine species analyzed show a trend for earlier spring arrival at Capri in most phases of migration (see table S4 online and Fig. 1). The long-term trend on Capri is at least as strong as that observed in Scandinavia. In short-distance migrants, instead, we find only a weak trend toward earlier arrival, and there is considerable variation between species (see tables S1 to S3 online and Fig. 1).

In accordance with previous findings (13–15),

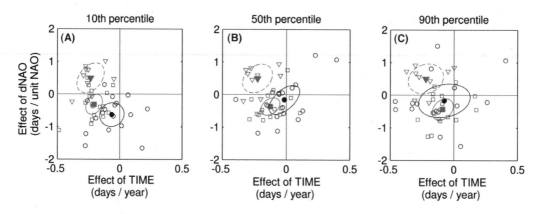

FIGURE 1. Long-term trend (TIME) and the effect of short-term climatic fluctuations (dNAO) on the early [(A), 10th percentile], mid [(B), 50th percentile], and late [(C), 90th percentile] phases of the spring arrival distribution in short-distance migrants (blue circles) and long-distance migrants (red squares) in Scandinavia and on Capri (red triangles). The solid symbols are sample averages, and the ellipses delimit their 95% confidence regions (11). Estimates for each species are given in tables S1 to S4. The differences in effect size for early-phase arrival of short-distance migrants versus long-distance migrants was 0.15 [95% confidence interval (CI): 0.06 to 0.23] days per year for the effect of TIME.

a high NAO index is associated with the early arrival of short-distance migrants in Scandinavia, but only in the early phase of migration (see Fig. 1). On the other hand, most long-distance migrants tend to arrive earlier in Scandinavia during years of high NAO in all phases of migration (see tables S1 to S3 online). The opposite pattern is observed at Capri, where high NAO tends to delay arrival times (see data in table S4 online). The underlying reason for this may be found south of the Sahara, because a high NAO index harms productivity over vast areas of northwestern and southeastern Africa (16), which may delay the spring departure of migrants from sub-Saharan wintering areas.

By showing that long-distance migrants have advanced their migration more than short-distance migrants, we have challenged the conventional wisdom that species wintering in temperate Europe should respond more strongly to climate change than trans-Saharan migrants (10). Furthermore, the earlier arrival of trans-Saharan migrants at Capri shows that the temporal trend for earlier arrival in Scandinavia cannot be explained simply by faster migration

through Europe in response to a concomitant trend of increasing temperatures taking place within continental Europe (17). Instead it suggests that (i) the onset of migration has advanced or (ii) the speed of migration through Africa has increased. Both alternatives could be seen as phenotypic responses to trends in the African climate patterns that have a positive effect on the foraging conditions (18), thereby improving the birds' physical conditions, which in turn affects their timing of migration (19) and makes the migration (including flight and stopover) more efficient. A positive trend in African temperatures (20) has previously been suggested as a reason why long-distance migrants arrive earlier in northern Europe (21). However, increasing African temperatures should decrease productivity (22), thereby delaying long-distance migrants' departure from the wintering ground. Hence, the earlier arrival is probably not a phenotypic response to improved foraging conditions. More likely, the rapid advance in arrival dates of long-distance migrants in Europe is due to climate-driven evolutionary changes in the timing of spring migration. Even though

SCIENCE IN THE NEWS

*Science*NOW Daily News, 22 February 2007

BRIGHT NIGHTS DIM SURVIVAL CHANCES
Phil Berardelli

WASHINGTON, D.C.—At a conference here yesterday, researchers reported that even low levels of light from incandescent, fluorescent, or other human-made sources can befuddle creatures that require a period of nighttime darkness. The findings add to the evidence that artificial lighting is interfering with the development, reproduction, and survival of species across the taxonomic spectrum.

All animals—from one-celled critters to humans—produce melatonin, a hormone that regulates cell metabolism, protects against the formation of cancerous tumors in larger animals, and allows many mammals and humans to enjoy restful sleep. But the hormone accumulates most efficiently in recurring or total darkness, such as in regular day-night cycles. When those cycles are disrupted, so is melatonin production. On the behavioral side, even seeing artificial illumination—such as streetlights or indoor lamps shining through windows—at night can throw off foraging and migration in many species.

To find out how brighter nights are altering metabolism and reproduction, herpetologist Bryant Buchanan of Utica College in New York and colleagues exposed snails and larval frogs to different levels of artificial light over periods lasting up to two months. With even the slightest amount of artificial light, the percentage of frogs developing normally dropped as low as 10%, compared with about 40% under more natural lighting conditions. The snail experiments produced similar results. Artificial illumination appears to produce "a dose response, not an on-off switch," Buchanan says. Constant lighting at night also suppressed the frogs' normal calling behavior and kept the snails hiding under leaf litter instead of searching for food.

Buchanan's findings are consistent with results for other species, says ecologist Travis Longcore of The Urban Wildlands Group in Los Angeles, California. "The introduction of light—even light that we would consider dim—will disrupt the natural cycles of animals, including humans," he says. An overlooked problem, he adds, is that outdoor lighting can hamper attempts to protect endangered wildlife living in or near urban areas. Longcore says he knows of one species of snake that disappeared from an urban habitat specifically set aside for it after steady levels of artificial light apparently disrupted its predation patterns, by exposing it either to its prey or to its own predators. "If we don't take [lighting effects] into account," he says, "our best-laid conservation plans will not succeed."

migratory activity is under endogenous control, experiments have demonstrated individual variation in the response to the photoperiodic cues needed to trigger the mechanisms underlying the onset of migration (23). The passerine birds investigated here reproduce at just one year of age and thus have the potential for a rapid evolutionary response to environmental changes. Given the considerable heritable genetic variation in the timing of migration (24, 25) and the selection pressure to breed earlier in Europe (6, 7), a change toward earlier arrival is indeed to be expected (26).

References and Notes

1. R. Harrington, I. Woiwood, T. H. Sparks, *Trends Ecol. Evol.* 14, 146 (1999).

2. T. L. Root *et al.*, *Nature* 421, 57 (2003).

3. D. Lack, *Ecological Adaptations for Breeding in Birds* (Methuen, London, 1968).

4. M. E. Visser, C. Both, M. M. Lambrecht, *Adv. Ecol. Res.* 35, 89 (2004).

5. P. O. Dunn, *Adv. Ecol. Res.* 35, 69 (2004).

6. C. Both, M. E. Visser, *Nature* 411, 296 (2001).

7. C. Both, S. Bouwhuis, C. M. Lessels, M. E. Visser, *Nature* 441, 81 (2006).

8. P. Berthold, *Control of Bird Migration* (Cambridge Univ. Press, Cambridge, 1996).

9. E. Gwinner, *J. Exp. Biol.* 199, 39 (1996).

10. E. Lehikoinen, T. H. Sparks, M. Zalakevicius, *Adv. Ecol. Res.* 35, 1 (2004).

11. The details of the data and methods are available as supporting material on *Science* Online.

12. J. W. Hurrell, *Science* 269, 676 (1995).

13. O. Hüppop, K. Hüppop, *Proc. R. Soc. London Ser. B* 270, 233 (2003).

14. M. Stervander, Å. Lindström, N. Jonzén, A. Andersson, *J. Avian Biol.* 36, 210 (2005).

15. A. V. Vähätalo, K. Rainio, A. Lehikoinen, E. Lehikoinen, *J. Avian Biol.* 35, 210 (2004).

16. L. C. Stige *et al.*, *Proc. Natl. Acad. Sci. USA* 103, 3049 (2006).

17. C. Both, R. G. Bijlsma, M. E. Visser, *J. Avian Biol.* 36, 368 (2005).

18. N. Saino *et al.*, *Ecol. Lett.* 7, 21 (2004).

19. P. P. Marra, K. A. Hobson, R. T. Holmes, *Science* 282, 1884 (1998).

20. M. Hulme, R. Doherty, T. Ngara, M. New, D. Lister, *Clim. Res.* 17, 145 (2001).

21. P. A. Cotton, *Proc. Natl. Acad. Sci. USA* 100, 12219 (2003).

22. O. Gordo, L. Brotons, X. Ferrer, P. Comas, *Global Change Biol.* 11, 12 (2005).

23. T. Coppack, F. Pulido, M. Czisch, D. P. Auer, P. Berthold, *Proc. R. Soc. London Ser. B* 270, S43 (2003).

24. A. P. Møller, *Proc. R. Soc. London Ser. B* 268, 203 (2001).

25. F. Pulido, P. Berthold, in *Avian Migration*, P. Berthold, E. Gwinner, E. Sonnenschein, Eds. (Springer-Verlag, Berlin, 2003), pp. 53–77.

26. Funding for the analysis reported in this paper was provided by the Nordic Council through the NCoE-EcoClim and the Swedish Research Council (to N.J.). Funding for obtaining the Ottenby data was provided by the Swedish Environmental Protection Agency. This is contribution number 215 from Ottenby Bird Observatory, contribution number 232 from Falsterbo Bird Observatory, contribution number 79 from Jomfruland Bird Observatory, and results from the Progetto Piccole Isole (Istituto Nazionale per la Fauna Selvatica), paper no. 37.

Supporting Online Material

www.sciencemag.org/cgi/content/full/312/5782/1959/DC1
Methods
SOM Text
Tables S1 to S5
References

Perspectives on the Arctic's Shrinking Sea-Ice Cover

MARK C. SERREZE, MARIKA M. HOLLAND, JULIENNE STROEVE

L inear trends in arctic sea-ice extent over the period 1979 to 2006 are negative in every month. This ice loss is best viewed as a combination of strong natural variability in the coupled ice-ocean-atmosphere system and a growing radiative forcing associated with rising concentrations of atmospheric greenhouse gases, the latter supported by evidence of qualitative consistency between observed trends and those simulated by climate models over the same period. Although the large scatter between individual model simulations leads to much uncertainty as to when a seasonally ice-free Arctic Ocean might be realized, this transition to a new Arctic state may be rapid once the ice thins to a more vulnerable state. Loss of the ice cover is expected to affect the Arctic's freshwater system

---

This article first appeared in *Science* (16 March 2007: Vol. 315, no. 5818). It has been revised for this edition.

and surface energy budget and could be manifested in middle latitudes as altered patterns of atmospheric circulation and precipitation.

The most defining feature of the Arctic Ocean is its floating sea-ice cover, which has traditionally ranged from a maximum extent of about 16×10^6 km^2 in March to a minimum extent of 7×10^6 km^2 at the end of the summer melt season in September (Fig. 1). Consistent satellite-derived monthly time series of sea-ice extent are provided by the Nimbus-7 Scanning Multichannel Microwave Radiometer

(October 1978 to August 1987) and the Defense Meteorological Satellite Program Special Sensor Microwave/Imager (1987 to present). Based on regression analysis of the combined record over the period 1979 to 2006, ice extent has declined for every month (Fig. 2), most rapidly for September, for which the trend is −8.6 ± 2.9% per decade or about 100,000 km² per year. Ice extent is defined as the area of the ocean with a fractional ice cover (i.e., an ice concentration) of at least 15% (1–3).

Every year since 2001 has yielded pronounced September minima, the most extreme of which was in 2005 (5.56×10^6 km²). When compared with the mean ice extent over the period 1979 to 2000, this represents a spatial reduction of 21% (1.6×10^6 km²), an area roughly the size of Alaska (Fig. 1). Comparisons with earlier records, which combine visible-band satellite imagery and aircraft and ship reports, suggest that the September 2005 ice extent was the lowest in at least the past 50 years. Data for the past few years suggest an accelerating decline in winter sea-ice extent (4).

Evidence for accompanying reductions in ice thickness (5) is inconclusive. Upward-looking sonar aboard submarines provides information on ice draft—the component of the total thickness (about 90%) that projects below the water surface. Comparisons between early sonar records (1958 to 1976) and those for 1993 to 1997 indicate reductions of 1.3 m in mean late summer ice draft over much of the central Arctic Ocean (6), but sparse sampling complicates interpretation. Further analysis of the submarine-acquired data in conjunction with model simulations points to thinning through 1996 but modest recovery thereafter (7). Results from an ice-tracking algorithm applied to satellite data from 1978 to 2003 document decreasing coverage of old, thick ice (8).

Understanding the Observed Ice Loss

The observed decline in ice extent reflects a conflation of thermodynamic and dynamic processes. Thermodynamic processes involve changes in surface air temperature (SAT), radiative fluxes, and ocean conditions. Dynamic processes involve changes in ice circulation in response to winds and ocean currents. These include changes in the strength and location of the Beaufort Gyre (a mean annual clockwise motion in the western Arctic Ocean) and characteristics of the Transpolar Drift Stream (a motion of ice that progresses from the coast of Siberia, across the pole, and into the North Atlantic via the Fram Strait). Nearly all of the ice export from the Arctic to the Atlantic occurs through this narrow strait between northern Greenland and Svalbard (Fig. 1).

Estimated rates of change in SAT over the Arctic Ocean for the past several decades vary depending on the time period and season, as well as the data source being considered. Although natural variability plays a large role in SAT variations, the overall pattern is one of recent warming, which is, in turn, part of a global signal (9). Using a record that combined coastal station observations with data from drifting buoys (from 1979 onward) and Russian "North Pole" stations (1950 to 1991), Rigor *et al.* (10) found positive SAT trends from 1979 to 1997 that were most pronounced and widespread during spring. Although there are biases in the buoy data relative to the North Pole data, especially for October through April (11), independent evidence for warming during spring, summer, and autumn since 1981 is documented in clear-sky surface temperatures retrieved from advanced very high resolution radiometer satel-

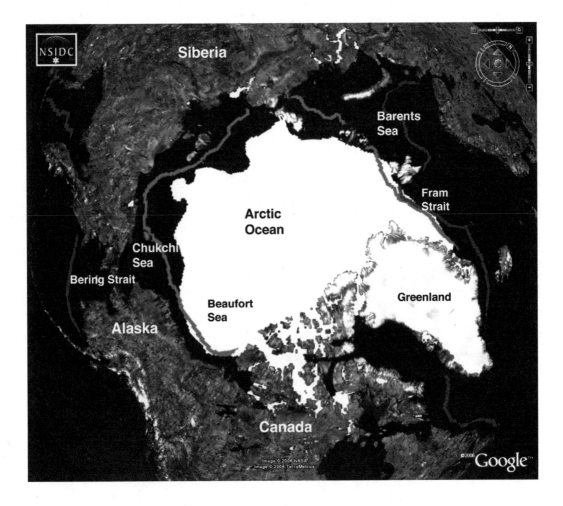

FIGURE 1. Sea-ice extent (bright white area) for September 2005. Median ice extents based on the period 1979 to 2000 for September (red line) and March (blue line) illustrate the typical seasonal range. Geographic features referred to in the text are labeled. Source: NSIDC image in Google Earth.

lite imagery (*12*). Further support for warming comes from analysis of satellite-derived passive microwave brightness temperatures that indicate earlier onset of spring melt and lengthening of the melt season (*13*), as well as from data from the Television Infrared Observation Satellites Operational Vertical Sounder that point to increased downwelling radiation to the surface in spring over the past decade, which is linked to increased cloud cover and water vapor (*14*). Our assessments of autumn and winter data fields from the National Centers for Environmental Prediction and National Center for Atmospheric Research (NCEP-NCAR) reanalysis (*15*) point to strong surface and low-level warming for the period 2000 to 2006 relative to 1979 to 1999. Weaker warming is evident for summer.

All of these results are consistent with a declining ice cover. However, at least part of the

KEY TERM

Downwelling: A downward flux of solar or long-wave radiation, that is, directed from the atmosphere toward the surface.

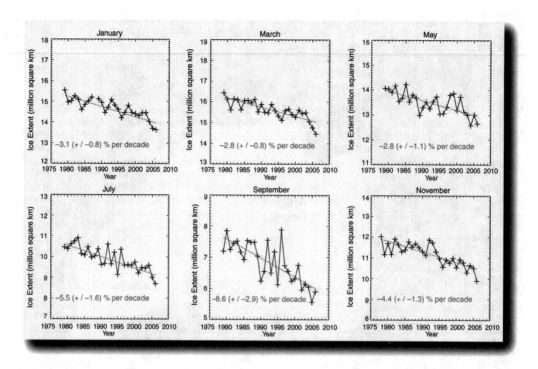

FIGURE 2. Time series of Arctic sea-ice extent for alternate months and least-squares linear fit based on satellite-derived passive microwave data from November 1979 through November 2006. Listed trends include (in parentheses) the 95% confidence interval of the slope. Ice extent is also declining for the six months that are not shown, ranging from −2.8 ± 0.8% per decade in February to −7.2 ± 2.3% per decade in August.

recent cold-season warming seen in the NCEP-NCAR data is itself driven by the loss of ice, because this loss allows for stronger heat fluxes from the ocean to the atmosphere. The warmer atmosphere then promotes a stronger longwave flux to the surface.

Links have also been established between ice loss and changes in ice circulation associated with the behavior of the North Atlantic Oscillation (NAO), Northern Annular Mode (NAM), and other atmospheric patterns. The NAO refers to covariability between the strength of the Icelandic Low and that of the Azores High, which are the two centers of action in the North Atlantic atmospheric circulation. When both are strong (or weak), the NAO is in its positive (or negative) phase. The NAM refers to an oscillation of atmospheric mass between the

arctic and middle latitudes and is positive when arctic pressures are low and mid-latitude pressures are high. The NAO and NAM are closely related and can be largely viewed as expressions of the same phenomenon.

From about 1970 through the mid-1990s, winter indices of the NAO-NAM shifted from negative to strongly positive. Rigor et al. (16) showed that altered surface winds resulted in a more cyclonic motion of ice and an enhanced transport of ice away from the Siberian and Alaskan coasts (i.e., a more pronounced Trans-polar Drift Stream). This change in circulation fostered openings in the ice cover. Although these openings quickly refroze in response to low winter SATs, coastal areas in spring were nevertheless left with an anomalous coverage of young, thin ice. This thin ice then melted out

*Science*NOW Daily News, 7 September 2007

DWINDLING DAYS FOR ARCTIC ICE
Phil Berardelli

If computer models are correct, by 2050, Arctic sea ice will shrink during late summer by more than twice as much as it does now. The results of a new study by researchers at the National Oceanic and Atmospheric Administration (NOAA) add weight to speculation that a northern sea route will open up from Europe to Asia for the first time in recorded history.

The Arctic ice cap remains one of the most variable features of our planet. For many millennia, the frozen areas of the Northern Hemisphere have advanced and retreated, while a similar but smaller variation occurs on a seasonal basis. Now, between ice ages and in the midst of an upward trend in average temperatures, the Arctic Ocean's ice is showing signs of unprecedented summer shrinkage. That development could be a boon to international ship commerce but a potentially serious threat to the ecosystems that have emerged within polar environments. Furthermore, the transformation of the reflective white ice into heat-absorbing seawater could further accelerate the warming of the planet.

To carry out their study, oceanographer James Overland of NOAA's Pacific Marine Environmental Laboratory and meteorologist Muyin Wang of the agency's Joint Institute for the Study of the Atmosphere and Ocean at the University of Washington, both in Seattle, selected 11 climate models that closely predicted actual amounts of Arctic ice area from 1979 to 1999. Then they directed the computer programs to look ahead to 2050. The result, the team reports in the 8 September 2007 issue of *Geophysical Research Letters*, is that summer Arctic sea-ice area could shrink by more than 40% and could open waters off Alaska, Canada, and Russia that historically have remained icebound. That compares with about 18% summer shrinkage, on average, from 1979 to 1999. The models also project less ice formation during the winter in the Bering and Barents Seas and in the Sea of Okhotsk, although not in Canada's Baffin Bay.

The 40% figure could be conservative, says ice scientist Waleed Abdalati of NASA's Goddard Space Flight Center in Greenbelt, Maryland, "as even the best models have historically underestimated the current rate of ice decline." But one thing is clear, he says: "The dramatic losses we are seeing in Arctic ice cover are not expected to slow down." Research scientist Jianli Chen of the University of Texas, Austin, thinks that the planet will experience a "snowball effect." Shrinking sea ice will increase the ocean's heat absorption, he says, which will in turn "further increase the melting of sea ice and contribute to global warming."

in summer, which was expressed as large reductions in ice extent. Summer ice loss was further enhanced as the thinner ice promoted stronger heat fluxes to the atmosphere, fostering higher spring air temperatures and earlier melt onset.

Given that the NAO-NAM has regressed back to a more neutral state since the late 1990s (17), these processes cannot readily explain the extreme September sea-ice minima of recent years. Rigor and Wallace (18) argued that recent extremes represent delayed impacts of the very strongly positive winter NAO-NAM state from about 1989 to 1995. As the NAO-NAM rose to this positive state, shifts in the wind field not only promoted the production of thinner spring ice in coastal areas but flushed much of the Arctic's store of thick ice into the North Atlantic through Fram Strait.

Rothrock and Zhang (19) modified this view. Using a coupled ice-ocean model, they argued that although wind forcing was the dominant driver of declining ice thickness and volume from the late 1980s through mid-1990s, the ice response to generally rising air temperatures was more steadily downward over the study period (1948 to 1999). In other words, without the NAO-NAM forcing, there would still have been a downward trend in ice extent, albeit smaller than that observed. Lindsay and Zhang (20) came to similar conclusions in their modeling study. Rising air temperature has reduced ice thickness, but changes in circulation also flushed some of the thicker ice out of the Arctic, leading to more open water in summer and stronger absorption of solar radiation in the upper (shallower depths of the) ocean. With more heat in the ocean, thinner ice grows in autumn and winter.

Recent years have experienced patterns of atmospheric circulation in spring and summer favoring ice loss. By altering both the Beaufort Gyre and Transpolar Drift Stream, these patterns have reduced how long ice is sequestered and aged in the Arctic Ocean (21). The strength of a cyclonic atmospheric regime that sets up over the central Arctic Ocean in summer is

KEY TERM

Cyclonic atmospheric regime: a period in which the atmospheric circulation in a given region is characterized by cyclonic flow (counterclockwise in the Northern Hemisphere).

important. Along with promoting offshore ice motion, the pronounced cyclonic summer circulations of 2002 and 2003 favored ice divergence, as is evident from the low ice concentrations in satellite imagery. Ice divergence in summer spreads the existing ice over a larger area, but enhanced absorption of solar energy in the areas of open water promotes stronger melt. There was also very little September ice in the Greenland Sea (off the east coast of Greenland) for these summers, which may also be linked to winds associated with this summer atmospheric pattern (22).

To further complicate the picture, it appears that changes in ocean heat transport have played a role. Warm Atlantic waters enter the Arctic Ocean through eastern Fram Strait and the Barents Sea and form an intermediate layer as they subduct below colder, fresher (less dense) Arctic surface waters. Hydrographic data show increased import of Atlantic-derived waters in the early to mid-1990s and warming of this inflow (23). This trend has continued, characterized by pronounced pulses of warm inflow. Strong ocean warming in the Eurasian Basin in 2004 can be traced to a pulse entering the Barents Sea in 1997 and 1998. The most recent data show another warm anomaly poised to enter the Arctic Ocean (24, 25). These inflows may promote ice melt and discourage ice growth along the Atlantic ice margin. Once Atlantic water enters the Arctic Ocean, the cold halocline layer (CHL) separating the Atlantic and surface waters largely insulates the ice from the heat of the Atlantic layer. Observations suggest a retreat of the CHL in the Eurasian Basin in the 1990s (26). This likely increased Atlantic layer heat loss and ice-ocean heat exchange. Partial recovery of the CHL has been observed since 1998 (27).

Maslowski et al. (28) proposed a connection between ice loss and oceanic heat flux through the Bering Strait. However, hydrographic data collected between 1990 and 2004 document strong variability in this inflow, as opposed to a longer-term trend. An observed increase in the flux between 2001 and 2004 is estimated to be capable of melting 640,000 km^2 of ice 1 m thick, but fluxes in 2001 are the lowest of the record (29). Subsequent analysis (30) nevertheless reveals a link between ice loss and increases in Pacific Surface Water (PSW) temperature in the Arctic Ocean beginning in the late 1990s, concurrent with the onset of sharp sea-ice reductions in the Chukchi and Beaufort seas. The hypothesis that has emerged from those observations is that delayed winter ice formation allows for more efficient coupling between the ocean and wind forcing. This redirects PSW from the shelf slope along Alaska into the Arctic Ocean, where it is more efficient in retarding winter ice growth. An imbalance between winter ice growth and summer melt results, accelerating ice loss over a large area.

To summarize, the observed sea-ice loss can in part be connected to Arctic warming over the past several decades. Although this warming is part of a global signal suggesting a link with greenhouse gases (GHG) loading, attribution is complicated by a suite of contributing atmospheric and oceanic forcings. Below we review the evidence for an impact of GHG loading on the observed trends and projections for the future, based on climate-model simulations.

Simulations from Climate Models

Zhang and Walsh (31) showed that most of the models used in the Intergovernmental Panel on Climate Change Fourth Assessment Report (IPCC AR4) have climatological sea-ice extent within 20% of the observed climatology over

their adopted base period of 1979 to 1999, with good simulation of the seasonal cycle. The multimodel ensemble mean realistically estimates observed ice-extent changes over this base period, and most individual models also show a downward trend. Our analysis of an IPCC AR4 multimodel ensemble mean hindcast (in which known or closely estimated inputs for past events are entered into the model to see how well the output matches the known results) for the longer base period 1979 to 2006 also reveals consistency with observations regarding larger trends in September versus those in winter. These results provide strong evidence that despite prominent contributions of natural variability in the observed record, GHG loading has played a role.

Rates of ice loss both for the past few decades and projected through the 21st century nevertheless vary widely between individual models. Our analyses show that in the IPCC AR4 models driven with the Special Report on Emissions Scenarios (SRES) A1B emissions scenario (in which atmospheric carbon dioxide reaches 720 parts per million by 2100), a near-complete or complete loss (to less than 1×10^6 km^2) of September ice will occur anywhere from 2040 to well beyond the year 2100, depending on the model and the particular run for that model. Overall, about half the models reach September ice-free conditions by 2100 (32). Figure 3 shows the spatial pattern of the percent of models that predict at least 15% fractional ice cover for March and September, averaging output over the period 2075–2084. Even by the late 21st century, most models project a thin ice cover in March. By contrast, about 40% of the models project no ice in September over the central Arctic Ocean.

The scatter among models reflects many factors, including the initial (late-20th century) simulated ice state, aspects of the modeled ocean circulation, simulated cloud conditions, and natural variability in the modeled system (e.g., NAO-NAM–like behavior). These tie in strongly to the strength and characteristics of the positive ice-albedo feedback mechanism. In general, GHG loading results in a stronger and longer summer melt season, thinning the ice and exposing more of the dark (low albedo) ocean surface that readily absorbs solar radiation. Autumn ice growth is delayed, resulting in thinner spring ice. This thin ice is more apt to melt out during the next summer, exposing more open water, which results in even thinner ice during the following spring. Negative feedbacks, such as the fact that thinner ice grows more rapidly than thicker ice when exposed to the same forcing, can counteract these changes but are generally weaker.

Although there is ample uncertainty regarding when a seasonally ice-free Arctic Ocean will be realized, the more interesting question is how it arrives at that state. Simulations based on the Community Climate System Model version 3 (CCSM3) (33) indicate that end-of-summer ice extent is sensitive to ice thickness in spring. If the ice thins to a more vulnerable state, a "kick" associated with natural climate variability can result in rapid summer ice loss because of the

ice-albedo feedback. In the events simulated by CCSM3, anomalous ocean heat transport acts as this trigger. Such abrupt transitions are typically four times as fast as the observed trends over the satellite record. In one ensemble member, September ice extent decreases from about $6 \times 10^6 \, \text{km}^2$ to $2 \times 10^6 \, \text{km}^2$ in 10 years, resulting in near ice-free September conditions by 2040. A number of other climate models show similar rapid ice-loss events.

Impacts

Loss of the sea-ice cover will have numerous impacts. A sharply warmer Arctic in autumn and winter is expected as a result of larger heat fluxes from the ocean to the atmosphere. This is the primary fingerprint of Arctic amplification of greenhouse warming (34). As ice retreats from the shore, winds have a longer fetch over open water, resulting in more wave action. This effect is already resulting in coastal erosion in Alaska and Siberia. Ice loss is also affecting traditional hunting practices by members of indigenous

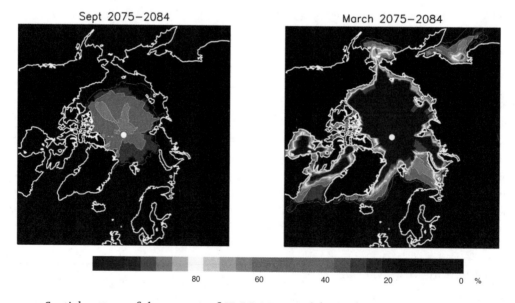

FIG. 3. Spatial pattern of the percent of IPCC AR4 model simulations (SRES A1B scenario) with at least 15% ice concentration for March (top) and September (bottom), averaged over the decade 2075 to 2084. For example, a value of 60% at a given location means that 60% of simulations predicted sea ice. Results are based on 11 models with realistic 20th-century September sea-ice extent.

cultures and contributing to regional declines in polar bear health and abundance (35).

In their modeling study, Magnusdottir et al. (36) found that declining ice in the Atlantic sector promotes a negative NAO-NAM atmospheric circulation response, with a weaker, southward-shifted storm track. Singarayer et al. (37) forced the Hadley Centre Atmospheric Model with observed sea ice from 1980 to 2000 and projected sea-ice reductions until 2100. In one simulation, mid-latitude storm tracks were intensified, increasing precipitation over western and southern Europe in winter. Experiments by Sewall and Sloan (38) revealed impacts on extrapolar precipitation patterns leading to reduced rainfall in the American West. Although results from different experiments with different designs vary, the common thread is that sea ice matters.

Climate models also indicate that by increasing upper-ocean stability and suppressing deepwater formation, North Atlantic freshening may disrupt the global thermohaline circulation, possibly with far-reaching consequences. Increased freshwater export from the Arctic is a potential source of such freshening. Observations implicate an Arctic source for freshening in the North Atlantic since the 1960s (39). Total freshwater output to the North Atlantic is projected to increase through the 21st century, with decreases in ice export more than compensated by the liquid freshwater export. However, reductions in ice melt and associated freshening in the Greenland-Iceland-Norwegian (GIN) seas resulting from a smaller ice transport through Fram Strait may more directly affect the deepwater formation regions and counteract increased ocean stability due to the warming climate (i.e., a warmer upper ocean is more stable). This outcome could help maintain deepwater formation in the GIN seas (40).

Conclusions

Natural variability, such as that associated with the NAO-NAM and other circulation patterns,

has and will continue to have strong impacts on the Arctic sea-ice cover. However, the observed ice loss for the Arctic Ocean as a whole, including the larger trend for September as compared with that of winter, is qualitatively reproduced in ensemble mean-climate-model hindcasts forced with the observed rise in GHG concentrations. This strongly suggests a human influence (31). However, there is a large amount of scatter between individual simulations, which contributes to uncertainty regarding rates of ice loss through the 21st century. An emerging issue is how a seasonally ice-free Arctic Ocean may be realized: will it result from a gradual decline with strong imprints of natural variability, or could the transition be rapid once the ice thins to a more vulnerable state? Links between altered ocean heat transport and observed ice loss remain to be resolved, as does the attribution of these transport changes, but pulses such as those currently poised to enter the Arctic Ocean from the Atlantic could provide a trigger for a rapid transition.

In this regard, future behavior of the CHL, which insulates the sea ice from the warm Atlantic layer, is a key wild card. Another uncertainty is the behavior of the NAO-NAM. Despite its return to a more neutral phase, there is evidence, albeit controversial, that external forcing may favor the positive state that promotes ice loss. The mechanisms are varied but in part revolve around the idea that stratospheric cooling, in response to increasing GHG concentrations or through ozone destruction, may "spin up" the polar stratospheric vortex (the cyclonic circulation of the stratosphere, which on average is approximately symmetric about the poles of both hemispheres), resulting in lower Arctic surface pressures. Another view is that the NAO-NAM could be bumped to a preferred positive state via warming of the tropical oceans (41). However, as noted earlier, declining sea ice in the Atlantic sector may invoke a negative NAO-NAM response (36).

Given the agreement between models and observations, a transition to a seasonally ice-free Arctic Ocean as the system warms seems increasingly certain. The unresolved questions regard when this new Arctic state will be realized, how rapid the transition will be, and what will be the impacts of this new state on the Arctic and the rest of the globe (42).

References and Notes

1. Ice extent time series are available from the National Snow and Ice Data Center (NSIDC) based on the application of the NASA team algorithm (used here) and a bootstrap algorithm to the passive microwave brightness temperatures (http://nsidc.org/data/seaice/). Trends computed from both are negative in all months, but those from the bootstrap series are slightly smaller (which yielded a September trend of −7.9% per decade). Trends are computed from anomalies referenced to means over the period 1979 to 2000. Surface melt in summer contaminates the passive microwave signal, resulting in the underestimation of ice concentration. Use of ice extent (a binary ice/no-ice classification) largely circumvents this problem.

2. Trends for all months are significant at the 99% confidence level, based on an F test with the null hypothesis of a zero trend. Trends are also significant (exceeding the 95% level) based on the approach of Weatherhead et al. (3), which computes the trend significance from the variance and autocorrelation of the residuals.

3. E. C. Weatherhead et al., J. Geophys. Res. 103, 10.1029/98JD00995 (1998).

4. J. C. Comiso, Geophys. Res. Lett. 33, L18504 (2006).

5. Ice thickness can be described from a probability distribution, which has a peak at about 3 m. Although ice at the peak of the distribution is predominantly multiyear ice that has survived one or more melt seasons and thicker than younger first-year ice (representing a single year's growth), ridging can result in very thick first-year ice (up to 20 to 30 m).

6. D. A. Rothrock, Y. Yu, G. A. Maykut, Geophys. Res. Lett. 26, 3469 (1999).

7. D. A. Rothrock, J. Zhang, Y. Yu, J. Geophys. Res. 108, 3083 (2003).

8. C. Fowler, W. J. Emery, J. A. Maslanik, IEEE Geosci. Remote Sens. Lett. 1, 71 (2004).

9. M. C. Serreze, J. A. Francis, Clim. Change 76, 241 (2006).

10. I. G. Rigor, R. L. Colony, S. Martin, J. Clim. 13, 896 (2000).

11. I. V. Polyakov et al., J. Clim. 16, 2067 (2003).

12. J. C. Comiso, J. Clim. 16, 3498 (2003).

13. J. C. Stroeve, T. Markus, W. N. Meier, Ann. Glaciol. 25, 382 (2006).

14. J. A. Francis, E. Hunter, EOS Trans. Am. Geophys. Union 87, 509 (2006).

15. E. Kalnay et al., Bull. Am. Meteorol. Soc. 77, 437 (1996).

16. I. G. Rigor, J. M. Wallace, R. L. Colony, J. Clim. 15, 2648 (2002).

17. J. E. Overland, M. Wang, Geophys. Res. Lett. 32, L06701 (2005).

18. I. G. Rigor, J. M. Wallace, Geophys. Res. Lett. 31, L09401 (2004).

19. D. A. Rothrock, J. Zhang, J. Geophys. Res. 110, C01002 (2005).

20. R. W. Lindsay, J. Zhang, J. Clim. 18, 4879 (2005).

21. J. A. Maslanik, S. Drobot, C. Fowler, W. Emery, R. Barry, Geophys. Res. Lett. 34, 10.1029/2006GL028269 (2007).

22. J. C. Stroeve et al., Geophys. Res. Lett. 32, L04501 (2005).

23. R. R. Dickson et al., J. Clim. 13, 2671 (2000).

24. I. V. Polyakov et al., Geophys. Res. Lett. 32, L17605 (2005).

25. W. Walczowski, J. Piechura, Geophys. Res. Lett. 33, L12601 (2006).

26. M. Steele, T. J. Boyd, J. Geophys. Res. 103, 10419 (1998).

27. T. J. Boyd, M. Steele, R. D. Muench, J. T. Gunn, Geophys. Res. Lett. 29, 1657 (2002).

28. W. Maslowski, D. C. Marble, W. Walczowski, A. J. Semtner, Ann. Glaciol. 33, 545 (2001).

29. R. A. Woodgate, K. Aagaard, T. L. Weingartner, Geophys. Res. Lett. 33, L15609 (2006).

30. K. Shimada et al., Geophys. Res. Lett. 33, L08605 (2006).

31. X. Zhang, J. E. Walsh, J. Clim. 19, 1730 (2006).

32. O. Arzel, T. Fichefet, H. Goosse, Ocean Model. 12, 401 (2006).

33. M. M. Holland, C. M. Bitz, B. Tremblay, Geophys. Res. Lett. 33, L23503 (2006).

34. S. Manabe, R. J. Stouffer, J. Geophys. Res. 85, 5529 (1980).

35. I. Stirling, C. L. Parkinson, Arctic 59, 261 (2006).

36. G. Magnusdottir, C. Deser, R. Saravanan, J. Clim. 17, 857 (2004).

37. J. S. Singarayer, J. Bamber, P. J. Valdes, J. Clim. 19, 1109 (2006).

38. J. O. Sewall, L. C. Sloan, Geophys. Res. Lett. 31, L06209 (2004).

39. B. J. Peterson et al., Science 313, 1061 (2006).

40. M. M. Holland, J. Finnis, M. C. Serreze, J. Clim. 19, 6221 (2006).

41. N. P. Gillett, M. P. Baldwin, M. R. Allen, in The North Atlantic Oscillation: Climate Significance and Environmental Impact, J. W. Hurrell, Y. Kushnir, G. Ottersen, M. Visbeck, Eds., Geophysical Monograph Series 134 (American Geophysical Union, Washington, DC, 2003), chap. 9.

42. This study was supported by NSF, NASA, and NOAA. M. Savoie, L. Ballagh, W. Meier, and T. Scambos are thanked for their assistance.

Recent Sea Level Contributions of the Antarctic and Greenland Ice Sheets

ANDREW SHEPHERD AND DUNCAN WINGHAM

After a century of polar exploration, the past decade of satellite measurements has painted an altogether new picture of how Earth's ice sheets are changing. As global temperatures have risen, so have rates of snowfall, ice melting, and glacier flow. Although the balance between these opposing processes has varied considerably on a regional scale, data show that Antarctica and Greenland are each losing mass overall. Our best estimate of their combined imbalance is about 125 gigatons (Gt) of ice per year, enough to raise sea level by 0.35 mm per year. This is only a modest contribution to the present rate of sea level rise of 3.0 mm per year. However, much of the loss from Antarctica and Greenland is the result of the flow of ice to the ocean from ice streams and glaciers, which has accelerated over the past decade. In both continents, there are suspected triggers for the accelerated ice discharge—surface and ocean warming, respectively—and, over the course of the 21st century, these processes could rapidly counteract the snowfall gains predicted by present coupled climate models.

Antarctica and Greenland hold enough ice to raise global sea levels by some 70 m (1), and, according to the geological record (2), collapses of Earth's former ice sheets have caused increases of up to 20 m in less than 500 years. Such a rise, were it to occur today, would have tremendous societal implications (3). Even a much more gradual rise would have great impact. Accordingly, one goal of glaciological survey [e.g., (4, 5)] is to determine the contemporary sea level contribution due to Antarctica and Greenland. For much of the 20th century, however, the size of these ice sheets hindered attempts to constrain their mass trends, because estimating whole–ice sheet mass change could

This article first appeared in *Science* (16 March 2007: Vol. 315, no. 5818). It has been revised for this edition.

be done only by combining sparse local surveys, with consequent uncertainty. For example, a 1992 review (6) concluded that the available glaciological measurements allowed Antarctica to be anything from a 600 Gt/year sink to a 500 Gt/year source of ocean mass [500 Gt of ice equals 1.4 mm equivalent sea level (ESL)], accounting for nearly all of the 20th-century sea level trend of 1.8 mm/year (1) or, in the other direction, leaving a mass shortfall of some 1000 Gt/year. Even the 2001 Intergovernmental Panel on Climate Change (IPCC) report (1) preferred models to observations in estimating Antarctic and Greenland sea level contributions.

However, in the past decade, our knowledge of the contemporary mass imbalances of Antarctica and Greenland has been transformed by the launch of a series of satellite-based sensors. Since 1998, there have been at least 14 satellite-based estimates (7–20) of the mass imbalance of Earth's ice sheets (Table 1). At face value, their range of some −366 to 53 Gt/year, or 1.0 to −0.15 mm/year sea level rise equivalent, explains much of the eustatic component of 20th-century sea level rise [1.5 mm/year in (21)], but we argue that the contribution is smaller and the problem of closing the 20th-century sea level budget remains. Equally, the new observations provide a picture of considerable regional variability and, in particular, the long-predicted [e.g., (1)] snowfall-driven growth [e.g., (10, 22)] is being offset by large mass losses from particular ice-stream and glacier flows [e.g., (12, 23)]. There is, moreover, evidence in Greenland and Antarctica of recent accelerations in these flows (12, 24, 25). It is apparent that the late 20th- and early 21st-century ice sheets at least are dominated

by regional behaviors that are not captured in the models on which the Intergovernmental Panel on Climate Change (IPCC) predictions have depended, and there is renewed speculation (26, 27) of accelerated sea level rise from the ice sheets under a constant rate of climate warming.

Although the observations in Table 1 have narrowed the uncertainty in estimates of the eustatic contribution to sea level, the range of values is notably wider than their stated uncertainties. Accordingly, we give consideration to the limitations of the three methods—accounting the mass budget [e.g., (9)], altimetry measurement of ice sheet volume change [e.g., (7)], and observing the ice sheets' changing gravitational attraction [e.g., (11)]—used to calculate the estimates in Table 1. In light of these limitations, we discuss the recent changes in the Antarctic and Greenland ice sheets, and we conclude with some remarks on the future evolution of the ice sheets.

Methods and Their Sensitivity to Accumulation Rate

The mass-budget method [e.g., (9, 12)] compares the mass gain due to snowfall with mass losses due to sublimation, meltwater runoff, and ice that flows into the ocean. It has been given new impetus by the capability of interferometric synthetic aperture radar (InSAR) to determine ice surface velocity. This has improved earlier estimates of the ice flux to the ocean (5) and provides a capability to identify accelerations of ice flow. The method is hampered by a lack of accurate accumulation and ice thickness data. For Antarctica, where surface melting is negligible, accumulation may be determined by spatially averaging the history of accumulation recorded in ice cores, or from meteorological forecast models. Estimates of the temporally averaged accumulation or "mean" accumulation range, respectively, from 1752 to 1924 Gt/year (1) and from 1475 to 2331 Gt/year (28). The meteorological data are acknowledged to be of inferior

TABLE 1. Mass balance (MB) of the East Antarctic (EAIS), West Antarctic (WAIS), Antarctic (AIS), and Greenland (GIS) ice sheets as determined by a range of techniques and studies. Not all studies surveyed all of the ice sheets, and the surveys were conducted over different periods within the time frame 1992 to 2006. For comparison, 360 Gt of ice is equivalent to 1 mm of eustatic sea-level rise.

Study	Survey period	Survey area 10^6 km^2 (%)	EAIS MB Gt year^{-1}	WAIS MB Gt year^{-1}	AIS MB Gt year^{-1}	GIS MB Gt year^{-1}
Wingham et al. (7)[*]	1992–1996	7.6 (54)	−1 ± 53	−59 ± 50	−60 ± 76	
Krabill et al. (8)[*]	1993–1999	1.7 (12)				−47
Rignot and Thomas (9)	1995–2000	7.2 (51)	22 ± 23	−48 ± 14	−26 ± 37	
Davis and Li (17)[*]	1992–2002	8.5 (60)			42 ± 23	
Davis et al. (10)[*]	1992–2003	7.1 (50)	45 ± 7			
Velicogna and Wahr (11)	2002–2004	1.7 (12)				−75 ± 21
Zwally et al. (18)[*]	1992–2002	11.1 (77)	16 ± 11	−47 ± 4	−31 ± 12	11 ± 3
	1996					−83 ± 28
Rignot and Kanagaratnam (12)	2000	1.2 (9)				−127 ± 28
	2005					−205 ± 38
Velicogna and Wahr (20)	2002–2005	12.4 (88)	0 ± 51	−136 ± 19	−139 ± 73	
Ramillien et al. (19)	2002–2005	14.1 (100)	67 ± 28	−107 ± 23	−129 ± 15	−169 ± 66
Wingham et al. (14)[*]	1992–2003	8.5 (60)			27 ± 29	
Velicogna and Wahr (13)	2002–2006	1.7 (12)				−227 ± 33
Chen et al. (15)	2002–2005	1.7 (12)				−219 ± 21
Luthcke et al. (16)	2003–2005	1.7 (12)				−101 ± 16
Range			−1 to 67	−136 to −47	−139 to 42	−227 to 11

[*] Altimetry.
InSAR mass budget.
Gravimetry.

accuracy (28), and their wide range can perhaps be discounted. The range of the core-based estimates, which use substantially the same core records, arises from differences in their **spatial interpolation**. Recent compilations have used the satellite-observed microwave temperature, which is correlated with accumulation, to guide the interpolation, and a careful study (29) placed the error of individual drainage-basin accumulation at 5%. The extent to which this error may average out over the entire sheet is not known: The microwave interpolation field [see (29)] depends on factors other than accumulation (e.g., temperature) that may bias the outcome. There is also the difficulty that although accumulation is averaged over decades or centuries, the ice-flux measurements are limited to those of the satellite measurements [1995 to 2000 in (9)]. This complicates comparison of the estimated mass imbalance with altimetry estimates whose interval [e.g., 1992 to 2003 in (14)] is precisely defined. To date, 58% of Antarctica has been surveyed, although the method may in principle be extended to the remainder. Some 70% of Greenland has also been surveyed, but the impact of the satellite observations in determining the time-averaged imbalance is lessened because runoff from land-terminated ice, which in Greenland accounts for some 60% of the mass loss, remains largely unmeasured.

KEY TERM

Spatial interpolation is the procedure of estimating the value of properties at unsampled sites within an area bounded by existing observations.

The range of estimates of net accumulation and runoff [169 to 283 Gt/year in (1), about 20% of the total accumulation] has complicated mass imbalance estimates for some time [e.g., (30, 31)] and will continue to do so.

Satellite and aircraft, radar, and laser altimetry provide a detailed pattern of change in the ice sheets' interiors (7, 10, 14, 17, 18) and have played a key role in distinguishing changes related to accumulation and ice dynamics. The longest records to date span 1992 to 2003 (10, 14), and imbalances estimated from them differ from longer averages estimated by other methods as a result of fluctuations in accumulation and ablation. Ice cores [see (7)] and model reanalyses (28, 32) show fluctuations in accumulation, relative to their temporal means, on the order of 15% in individual years, and a similar variability in rates of ablation (32, 33) (Fig. 1). The problem is exacerbated because the density of snow differs from that of ice by a factor of three, and decadal fluctuations in snowfall mass are exaggerated in the observed volume fluctuations over those due to ice dynamics in the same ratio. A correction is possible if the snowfall fluctuation is independently known, but the only estimates available today are from meteorological forecast models, and a recent study (14) of Antarctica concluded that there was too little correspondence between the altimeter and meteorological data sets for this method to be reliable. Differences between estimates of mass change made from the same observations of volume change (10, 14, 18) arise largely through different approaches to the conversion of volume to mass. To give an idea of the uncertainty, Wingham *et al.* (14) showed that in the absence of other data, an altimeter estimate covering 73% of the Antarctic interior could vary by 90 Gt/year without contradicting the observed volume change. European Remote Sensing (ERS)–1 and ERS-2 radar altimetry (for which the longest records are available) has been limited to latitudes between 81.5°N and 81.5°S and to terrain of low slope. Because these regions lie in the ice sheet's interior, which is characterized by growth in Greenland in general and in

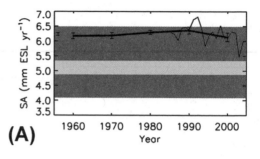

(A)

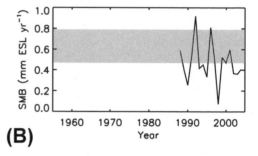

(B)

FIGURE 1. Fluctuations in (A) the rate of snow accumulation (SA) of Antarctica [redrawn from (38)] and (B) the net surface-mass balance (SMB) of Greenland [drawn from the data of (32)], determined from model reanalyses of meteorological observations expressed as ESL rise. Also shown are the ranges of published mean accumulation rates determined from glaciological observations (yellow) and climate models (blue).

Antarctica in some places, there is a tendency for these estimates to be more positive (Table 1). These difficulties may be overcome with the satellite laser altimeter records initiated by ICESat (Ice, Cloud, and land Elevation Satellite) in 2003 or, in the future, with the high-resolution radar altimeter of CryoSat-2.

Although differences in the time-averaged imbalances from the interferometric and altimetry methods are to be expected, the methods are highly complementary. For example, the retreat of the West Antarctic Pine Island Glacier grounding line observed by InSAR (34) is in close agreement with the drawdown of the inland ice observed with satellite altimetry (23). More recently, a combination of the two methods has provided considerable insight into

*Science*NOW Daily News, 15 February 2007

PREDICTING FATE OF GLACIERS PROVES SLIPPERY TASK

Richard A. Kerr

Earlier this month, the Intergovernmental Panel on Climate Change (IPCC) declined to extrapolate the recent accelerated loss of glacial ice far into the future. Too poorly understood, the IPCC authors said. Overly cautious, some scientists responded in very public complaints. The accelerated ice loss—apparently driven by global warming—could raise sea level much faster than the IPCC was predicting, they said. Yet almost immediately, new findings have emerged to support the IPCC's conservative stance.

In a surprise development, glaciologists reported online last week in *Science* that two major outlet glaciers draining the Greenland Ice Sheet—Kangerdlugssuaq and Helheim—did a lively two-step in the first part of the decade. By gauging the elevation and flow speed of the glaciers using satellite data, Ian Howat of the University of Washington's Applied Physics Laboratory in Seattle and his colleagues found that Kangerdlugssuaq sped up abruptly in 2005, no doubt accelerating sea level rise just a bit. But then it fell back to near its earlier flow speed by the next year. Helheim gradually accelerated over several years, also sped up sharply in 2005, and then slowed abruptly to its original flow speed. Apparently, these glaciers were temporarily responding to the loss of some restraining ice at their lower ends, much as a river's flow would temporarily increase with the lowering of a dam.

Helen Fricker of Scripps Institution of Oceanography in San Diego, California, and her colleagues report another glaciological surprise in a paper published online today in *Science*. Fricker also presented the study this morning at the annual meeting of the American Association for the Advancement of Science (which publishes *Science*NOW) in San Francisco, California. Using a new satellite-based laser technique, the team discovered an unexpectedly active network of linked lakes beneath two ice streams—Whillans and Mercer—draining the West Antarctic Ice Sheet. Researchers knew of pools of meltwater at the base of Antarctic ice, but Fricker and her colleagues recorded the rising and falling of the surface by up to 9 m over 14 patches of ice, the largest three spanning 120 to 500 km². Water that could lubricate the base of the ice and perhaps accelerate its flow was seeping from one subglacial lake to another in a matter of months, and in one case escaping to the sea. "We didn't know as much about the Antarctic Ice Sheet as we thought we did," says Fricker.

Glaciologist Richard Alley of Pennsylvania State University in State College agrees. "Lots of people were saying we [IPCC authors] should extrapolate into the future," he says, but "we dug our heels in at the IPCC and said we don't know enough to give an answer." Researchers will have to understand how and why glacier speeds can vary so much, he adds, before they can trust their models to forecast the fate of the ice sheets, much less sea level.

the unstable hydraulic connection between subglacial lakes in East Antarctica (35).

The GRACE (Gravity Recovery and Climate Experiment) satellites have permitted the changing gravitational attraction of the ice sheets to be estimated (11, 13, 15, 16, 19, 20). These estimates (Table 1) are more negative than those provided by mass budget or altimetry, but care is needed in making comparisons. The method is new, and a consensus about the measurement errors has yet to emerge [e.g., (36)], the correction for postglacial rebound is uncertain [e.g., (37)], contamination from ocean and atmosphere mass changes is possible [e.g., (16)], and the results depend on the method used to reduce the data [compare, e.g., (20) and (16)]. The GRACE record is also short (three years) and, as was the case with early altimeter time series [e.g., (7)], is particularly sensitive to the fluctuations in accumulation described above. For example, whereas (13) puts the total 2002 to 2005 Antarctic Ice Sheet mass loss at 417 ± 219 Gt, a subsequent meteorological study (38) has put the 2002 to 2003 snowfall deficit at 309 Gt, a value that explains most of the observed change.

East Antarctica

Although the East Antarctic Ice Sheet (EAIS) is the largest reservoir of ice on Earth, it exhibits the smallest range of variability among recent mass-balance estimates (Table 1). Since 1992, altimetric (7, 10, 14, 17, 18), interferometric (9), and gravimetric (13, 19) surveys have put the EAIS annual mass trend in the range −1 to 67 Gt/year. Growth of the EAIS mitigates the current sea level rise. Gains are limited to Dronning

Maud Land and Wilkes Land, and their spatial distribution (Fig. 2A) is strongly suggestive of snowfall-driven growth. Two glaciers in East Antarctica are losing mass (Fig. 2B). From 1992 to 2003, the fast-flowing trunks of the Totten and Cook glaciers deflated by 5.0 ± 0.5 km^3 per year and 2.4 ± 0.2 km^3 per year. Although these figures are only in rough coincidence with those determined from interferometry [0 ± 2 km^3 per year and −8 ± 5 km^3 per year, respectively, in (9)], the signals are clear and the trends definitely established.

West Antarctica and the Antarctic Peninsula

The West Antarctic Ice Sheet (WAIS) contains enough ice to raise global sea levels by more than 5 m and, according to altimetry and interferometry, one key sector is in a state of rapid retreat (23, 34). Glaciers draining into the Amundsen Sea (Fig. 2A) are losing mass because of an ice-dynamic perturbation. During the 1990s, for example, the Pine Island Glacier retreated by up to 1.2 km/year (34), thinned by up to 1.6 m/year (23), and accelerated by around 10% (39); the ice loss has been implicated in the freshening of the Ross Sea some 1000 km away (40). Throughout the 1990s, independent altimeter (7, 14, 17, 18) and interferometer (9) surveys of the WAIS as a whole were in notable, possibly fortuitous, agreement (Table 1), placing its

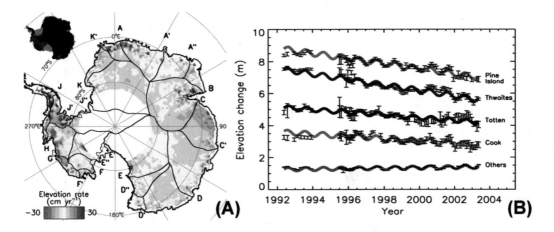

(A) (B)

FIG. 2. (A) Rate of elevation change of the Antarctic Ice Sheet, 1992 to 2003, from ERS satellite radar altimetry [redrawn from (14)]. Also shown (inset) is the bedrock geometry, highlighting floating (light gray), marine-based (mid-gray) and continental-based (black) sectors. (B) Elevation change of the trunks (flow in excess of 50 m/year) of the Pine Island [Basin GH in (A)], Thwaites (Basin GH), Totten (Basin C'D), and Cook (Basin DD') glaciers. All the deflating glaciers coincide with marine-based sectors of the ice sheet. An ice-dynamic origin of the thinning of the East Antarctic glaciers has yet to be confirmed by interferometry. However, the correlation of the thinning with flow velocity and the fact that the thinning rate is secular make ice dynamics the likely cause of all Antarctic mass losses.

annual losses in the range 47 to 59 Gt/year. The mass balance of the WAIS has been dominated by the losses from glaciers of the Amundsen sector, canceled to a degree by some snowfall-driven coastal growth and growth arising from the well-established shutdown of the Kamb Ice Stream (41).

There has been a report of an accelerated recent sea level contribution (42) based on satellite and aircraft altimetry, and the gravimetric surveys have also estimated a rate of mass loss since 2002 of between 107 and 136 Gt/year (Table 1). Such an acceleration (an increase in sea level trend of 0.2 mm/year, or about 10%) would be a cause for considerable concern. However, the altimeter data from which accelerated mass losses were derived in (42) span less than 5% of the WAIS area and use three altimeters with markedly different measurement errors. Furthermore, both data sets span a short time interval in which forecast models indicate that a 309 Gt accumulation deficit

occurred (38). Taking these factors into account, it is unlikely that the WAIS mass loss has altered substantially since the 1990s.

During the past decade, there have been notable glaciological changes at the Antarctic Peninsula (AP): The Larsen Ice Shelf thinned (43) and sections collapsed (44), accelerating ice discharge into the oceans by some 0.07 mm/year ESL rise (45). However, the majority of AP ice forms the continental ice cap of Dyer Plateau. This exhibits snowfall-driven growth (Fig. 2A) that is sufficient to cancel the accelerated flow from the Larsen-A and Larsen-B catchments. The AP contribution to sea level is negligible.

Greenland

Since the most recent IPCC report, there have been seven estimates of Greenland mass imbalance based on satellite altimetry (18), interferometry (12), and gravimetry (11, 15, 16, 19, 20). There is consensus that during the 1990s, the

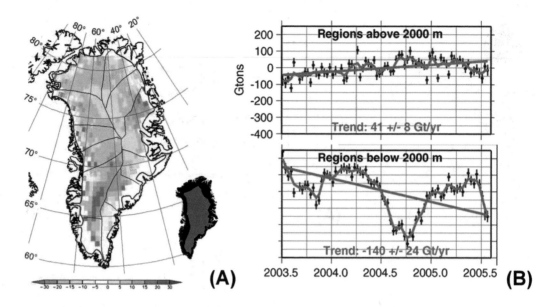

FIG. 3. (A) Rate of elevation change of the Greenland Ice Sheet, 1992 to 2003, determined from satellite radar altimetry [from (22)] and (B) time series of elevation change of individual sectors, 2003 to 2005, determined from satellite gravimetry [from (16)]. Also shown (inset) is the ice surface geometry, highlighting areas above (gray) and below (black) 2000 m elevation. Both instruments concur that high elevation areas are growing and low elevation areas are losing mass. According to gravimetry (16) and repeat InSAR measurements of ice discharge (12), the rate of mass loss at low elevations has increased over the past decade (see Table 1).

interior underwent modest snowfall-driven growth, which appears to be associated with a precipitation trend present in the meteorological record (32), offset by losses from lower altitude regions (Fig. 3, A and B). The decadal imbalance is not accurately determined. The more positive satellite altimeter estimate (18) is affected by signal loss in the steeper coastal margins; the aircraft laser measurements (8) are relatively sparse, although more sensitive to losses from marginal glaciers; and the mass-budget estimate (12) is undermined by the uncertainty of some 50 Gt/year in the accumulation. Nonetheless, the consensus of these measurements suggests a net loss in the 1990s of some 50 Gt/year.

Satellite interferometry (12, 24) has also established that from 1996 to 2005, mass losses through flow increased by 102 Gt/year, and meteorological estimates (32) (Fig. 1B) of the surface-mass imbalance decreased some 20 Gt/year in the same period because of increased melting. Gravimetric surveys too support an increased mass loss (11, 15, 16, 19, 20). However, the interferometric and gravimetric records are short and reflect the considerable variability in the mass flux of tidewater glaciers and the surface-mass balance. For example, the ice fluxes of two glaciers that by 2005 were responsible for 43 Gt/year of the increased discharge had by late 2006 declined to within 10 Gt/year of their level in 2000 (46), whereas in the past 14 years, the three-year variability in surface-mass imbalance has ranged from −130 to 120 Gt/year (Fig. 1B). In addition, not all gravimetric estimates capture the known spatial distribution of change; one that does (16) (Fig. 3B) is some 120 Gt/year more positive than other estimates, and some understanding of the cause of these discrepancies is needed. Increased mass loss from Greenland has occurred, but the decadal change is probably modest.

Implications for the Future

It is reasonable to conclude that, today, the EAIS is gaining some 25 Gt/year, the WAIS is losing about 50 Gt/year, and the Greenland Ice Sheet is losing about 100 Gt/year. These trends provide a sea level contribution of about 0.35 mm/year, a modest component of the present rate of sea level rise of 3.0 mm/year. Because 50 Gt/year is a very recent contribution, the ice sheets made little contribution to 20th-century sea level rise. However, what has also emerged is that the losses are dominated by ice dynamics. Whereas past assessments (47) considered the balance between accumulation and ablation, the satellite observations reveal that glacier accelerations of 20% to 100% have occurred over the past decade. The key question today is whether these accelerations may be sustained, or even increase, in the future.

The question is difficult because the causes of the instabilities have yet to be established. The geological record (48) suggests that some 10,000 years ago, the Amundsen sector of the WAIS extended only 100 km farther than today, confining the present rate of retreat to more recent times, and the drawdown of the Amundsen sector ice streams has been linked (49) to a recent trigger in the ocean. A comparable argument may be extended to the thinning glaciers in East Antarctica and Greenland, which are also marine terminated. Equally, there is no direct evidence of a warming of the Amundsen Sea, and it has long been held possible that the marine-terminated WAIS, and the Amundsen sector in particular, may be geometrically unstable (50), and the retreating East Antarctica streams have a similar geometry (Fig. 2A). In Greenland, where summer melting is widespread and increasing, Global Positioning System measurements have shown the melting to affect flow velocity in the ice sheet interior (26), introducing the possibility that increased surface meltwater is reaching the bed and accelerating the ice flow to the ocean.

The discovery that particular ice streams and glaciers are dominating ice sheet mass losses means that today our ability to predict future changes is limited. Present numerical models capture neither the details of actual ice streams nor, in Greenland, those of hydraulic connections between the surface and the bed. In addition, the detailed mechanics at the grounding line still remain to be fully worked out. In consequence, the view that the changing sea-level contribution of the Antarctic and Greenland ice sheets in the 21st century will be both small and negative as a result of accumulating snow in Antarctica [e.g., −0.05 mm/year in (1)] is now uncertain.

Because our predictive ability is limited, continued observation is essential. The satellite record clearly identifies the particular ice streams and glaciers whose evolution is of greatest concern. The causes of their instability need to be identified. Their detailed basal topography, their basal hydrology, and the details of the interaction with their surrounding shelf seas need to be established. Numerical models that capture the detailed dynamics of these glaciers and their hydrology are required. Of equal importance are meteorological and ice core measurements that will increase confidence in forecast models of accumulation and ablation fluctuations, because to a considerable extent, these limit interpretations of the short satellite records. There is a great deal that the International Polar Year may achieve.

References and Notes

1. J. A. Church, J. M. Gregory, in *Climate Change 2001: The Scientific Basis*, J. T. Houghton *et al.*, Eds. (Cambridge Univ. Press, Cambridge, 2001), chap. 11, pp. 641–693.

2. R. G. Fairbanks, *Nature* 342, 637 (1989).

3. N. Stern, *The Economics of Climate Change: The Stern Review* (Cambridge Univ. Press, Cambridge, 2006).

4. C. S. Benson, *Stratigraphic Studies in the Snow and Firn of Greenland Ice Sheet*, Research Report 70 (Cold Regions Research and Engineering Lab, Hanover, NH, 1962).

5. C. R. Bentley, M. B. Giovinetto, *Proceedings of the*

International Conference on the Role of Polar Regions in Global Change (Geophysical Institute, University of Alaska, Fairbanks, AK, 1991), pp. 481–486.

6. S. S. Jacobs, Nature 360, 29 (1992).

7. D. J. Wingham, A. Ridout, R. Scharroo, R. Arthern, C. K. Shum, Science 282, 456 (1998).

8. W. Krabill et al., Science 289, 428 (2000).

9. E. Rignot, R. H. Thomas, Science 297, 1502 (2002).

10. C. H. Davis, Y. Li, J. R. McConnell, M. M. Frey, E. Hanna, Science 308, 1898 (2005).

11. I. Velicogna, J. Wahr, Geophys. Res. Lett. 32, art-L18505 (2005).

12. E. Rignot, P. Kanagaratnam, Science 311, 986 (2006).

13. I. Velicogna, J. Wahr, Science 311, 1754 (2006).

14. D. J. Wingham, A. Shepherd, A. Muir, G. J. Marshall, Philos. Trans. R. Soc. A Math. Phys. Eng. Sci. 364, 1627 (2006).

15. J. L. Chen, C. R. Wilson, B. D. Tapley, Science 313, 1958 (2006).

16. S. B. Luthcke et al., Science 314, 1286 (2006).

17. C. H. Davis, Y. H. Li, Exploring and Managing a Changing Planet, paper presented at Science for Society (IEEE, Anchorage, Alaska, 20–24 September 2004), pp. 1152–1155.

18. H. J. Zwally et al., J. Glaciol. 51, 509 (2005).

19. G. Ramillien et al., Global Planet. Change 53, 198 (2006).

20. I. Velicogna, J. Wahr, Nature 443, 329 (2006).

21. W. Munk, Science 300, 2041 (2003).

22. O. M. Johannessen, K. Khvorostovsky, L. P. Bobylev, Science 310, 1013 (2005).

23. A. Shepherd, D. J. Wingham, J. A. D. Mansley, H. F. J. Corr, Science 291, 862 (2001).

24. I. Joughin, W. Abdalati, M. Fahnestock, Nature 432, 608 (2004).

25. A. Luckman, T. Murray, R. de Lange, E. Hanna, Geophys. Res. Lett. 33, art-L03503 (2006).

26. H. J. Zwally et al., Science 297, 218 (2002).

27. R. Bindschadler, Science 311, 1720 (2006); http://www.sciencemag.org/cgi/ijlink?linkType=ABST&journalCode=sci&resid=311/5768/1720.

28. A. J. Monaghan, D. H. Bromwich, S. H. Wang, Philos. Trans. R. Soc. A Math. Phys. Eng. Sci. 364, 1683 (2006).

29. R. J. Arthern, D. P. Winebrenner, D. G. Vaughan, J. Geophys. Res. Atmos. 111, D06107 (2006).

30. E. J. Rignot, S. P. Gogineni, W. B. Krabill, S. Ekholm, Science 276, 934 (1997).

31. N. Reeh, H. H. Thomsen, O. B. Olesen, W. Starzer, Science 278, 205 (1997).

32. J. E. Box et al., J. Clim. 19, 2783 (2006).

33. N. P. M. van Lipzig, E. van Meijgaard, J. Oerlemans, Int. J. Climatol. 22, 1197 (2002).

34. E. J. Rignot, Science 281, 549 (1998).

35. D. J. Wingham, M. J. Siegert, A. Shepherd, A. S. Muir, Nature 440, 1033 (2006).

36. M. Horwath, R. Dietrich, Geophys. Res. Lett. 33, art-L07502 (2005).

37. M. Nakada et al., Mar. Geol. 167, 85 (2000).

38. A. J. Monaghan et al., Science 313, 827 (2006).

39. I. Joughin, E. Rignot, C. E. Rosanova, B. K. Lucchitta, J. Bohlander, Geophys. Res. Lett. 30, 1706 (2003).

40. S. S. Jacobs, C. F. Giulivi, P. A. Mele, Science 297, 386 (2002).

41. S. Anandakrishnan, R. B. Alley, Geophys. Res. Lett. 24, 265 (1997).

42. R. Thomas et al., Science 306, 255 (2004).

43. A. Shepherd, D. Wingham, T. Payne, P. Skvarca, Science 302, 856 (2003); http://www.sciencemag.org/cgi/external_ref?access_num=14593176&link_type=MED.

44. H. Rott, P. Skvarca, T. Nagler, Science 271, 788 (1996).

45. E. Rignot et al., Geophys. Res. Lett. 31, art-L18401 (2004).

46. I. M. Howat, I. Joughin, T. A. Scambos, Science 315, 1559 (2007).

47. R. B. Alley, P. U. Clark, P. Huybrechts, I. Joughin, Science 310, 456 (2005).

48. A. L. Lowe, J. B. Anderson, Quat. Sci. Rev. 21, 1879 (2002).

49. A. J. Payne, A. Vieli, A. P. Shepherd, D. J. Wingham, E. Rignot, Geophys. Res. Lett. 31, art-L23401 (2004).

50. J. Weertman, J. Glaciol. 13, 3 (1974).

Projecting the Future

Introduction

DONALD KENNEDY

Much of what we know or expect about the future of climate change depends on assessments made by the Intergovernmental Panel on Climate Change. The IPCC, as scientists and policymakers around the world call it, is an organization with joint parentage from the United Nations Environmental Programme and the World Meteorological Organization. It operates through Working Groups (Working Group I focuses on science, for example, and III focuses on economic aspects.) The results of these efforts are eventually synthesized into a Summary for Policymakers, which takes into account—as one Working Group chair puts it—both science-push and policy-pull. This balance is important to the nature of IPCC's assessments: the organization does not do original research; rather, it is responsible for comprehensive evaluation of the work of scientists that appears in the peer-reviewed scientific literature. The outcome thus takes into account policy considerations as well as scientific ones.

The IPCC consensus, revealed in its most recent report, can be outlined as follows. Earth's greenhouse (a classic metaphor known and used for more than 100 years) has experienced an increase in average global temperature of about 0.7°C. This effect has been due to the addition of carbon dioxide (CO_2) and other "greenhouse gases" (oxides of nitrogen, methane, and chlorofluorocarbons). These changes have been brought about largely as the result of human activities: industrial combustion and the burning of forests, yielding CO_2, the most prominent and influential greenhouse gas. The current concentration of CO_2 in Earth's atmosphere averages about 385 parts per million by volume (ppmv), an increase over the preindustrial level of 280 ppmv.

This trend may be expected to increase as present rates of CO_2 emission (equivalent in 2004 to about 7400 million tons a year) continue. Various general circulation models (GCMs) are used to predict the relationship between increased concentrations of atmospheric greenhouse gases, usually expressed in CO_2 equivalents, and the future course of global warming. The consensus estimates yield a range between 2.5°C and 7.0°C for the value of the average global temperature in 2100, with an accompanying sea level rise between 20 and 80 cm.

What About the Future?

We launch a consideration of future projections with a News account of an IPCC report issued early in 2007. John Bohannon, a journalist who covers such issues for *Science*, covered a meeting held by Working Group III in Bangkok. Working Group III, as we noted above, deals with the economics of climate change, both the costs of its impacts and the costs of trying to slow climate change through mitigation measures like carbon sequestration, or programs to reduce emissions through the adoption of less carbon-intensive kinds of fuel, or the adoption of end-use efficiency standards. Bohannon gives a broad overview of the costs of lowering CO_2 emissions to various levels. He offers an excellent example of how Working Groups develop their own contributions to IPCC assessments by employing model estimates of future climate change. But more important for policy, he illustrates the differences of view within the political community on how economic impacts should be calculated. The United States is widely seen as having the staunchest opposition to paying for carbon mitigation. But this is starting to change.

Should We Trust the IPCC?

The IPCC has had enormous influence on the world's governments, beginning back in the early 1990s with its first assessments that illustrated strong effects on average global temperature as a result of the addition of greenhouse gases. Its accomplishments were recognized much later in its receipt of the Nobel Peace Prize for 2007, which was awarded to the IPPC along with Al Gore. The strong scientific consensus it represents has become a dominant force in thinking about the likely future of climate change. But that consensus has occasionally been criticized by scientists who otherwise agree with most IPCC findings. Their view is that the possibility of nonlinear, dynamic changes—such as major melting of ice sheets once the average global temperature rises above some hypothesized "threshold" value—should receive more consideration and may have been lost in the establishment of consensus. Michael Oppenheimer of Princeton and Brian O'Neill, Mort Webster, and Shardul Agrawala voice these concerns in their Policy Forum, "The Limits of Consensus." With special attention to sea level rise, they argue that the narrowing of uncertainty limited the IPCC's projection of the rise by 2100 to 18–59 cm—but left aside the possibility of increased contributions from dynamic processes in the Greenland and West Antarctic Ice Sheets.

The controversy over this critique was then played out in an exchange between an influential group of IPCC scientists led by Susan Solomon, co-chair of Working Group I. This was followed by a rejoinder from the Oppenheimer group. Our purpose in including this material is to underscore the differences between two approaches to envisioning the likely future—but also to demonstrate how scientific differences are resolved by published discourse. The differences in projected sea level rise between the parties here is rather small; they are in close agreement about the fact that climate change is under way, and they support the modeling and experimental approaches being undertaken to extend our knowledge. But they disagree on the way in which scientific conclusions are best expressed in communicating them to policymakers. The difference turns on whether "consensus" processes like the IPCC tend to produce agreement through negotiation of various positions, thus paying too little attention to outlier views. Those concerned with the future of climate change should examine this exchange carefully and reach their own decisions about the best way forward.

Rising Seas

Sea level rise is understandably controversial given its importance and complexity: future sea levels will depend not only on ocean temperature, but on contributions from melting of the world's great polar ice caps in Greenland and in Antarctica. The next two articles consider how the melting ice sheets will affect the level of the oceans. The first of these, by Bette Otto-Bliesner and a group of colleagues representing half a dozen different academic and government institutions, uses paleoclimatic evidence from a past warm period (130,000 and 116,000 years ago) to suggest that present ice sheet instability might cause future sea level rise to occur more rapidly than previously thought. A second paper, by Stefan Rahmstorf, uses much more recent empirical data on past relationships between average global temperature and sea level to conclude, as did Otto-Bliesner *et al.,* that we may need to anticipate a sea level rise of 1 m or more by 2100. These papers are in agreement even though they are based on the behavior of ice sheets at very different periods of geological time.

Food Security

As rising food prices and shortages spark hunger riots, researchers investigate how the changing climate might affect crops in some of the world's most vulnerable countries. In "Prioritizing Climate Change Adaptation Needs for Food Security in 2030," David Lobell and his colleagues use statistical crop models and climate projections to analyze the risks to staple crops like wheat, rice, and maize in 12 regions susceptible to widespread hunger. While their results are various, and there is some uncertainty in the models, Lobell *et al.* find that in many cases, food security is threatened in the near term by climate change. This study is a critical starting point; further research is needed if we are to prevent famine by understanding the risks to key crops in specific areas and by determining strategies for adaptation.

The Stern Report

A major new entry into the debate on climate change and how to manage it was introduced late in 2006 by a commission of the government of the United Kingdom chaired by Dr. Nicholas Stern. *The Economics of Climate Change: The Stern Review* was published by Cambridge University Press early in 2007. Students are urged to consult the Stern Report in the original; here we have included a summary. The Stern Report produced a significant disagreement among economists, much of it dealing with the use of near-zero discount rates in accounting for future welfare. In a critique published as a Policy Forum in *Science*, the Yale economist William Nordhaus argues that by setting this parameter as Stern did and by setting a term called "consumption elasticity" to a low value, Stern suggests that society is indifferent to inequality, that is, he assigns equal value to the welfare of future and present citizens. By choosing these parameters, Nordhaus argues, Stern makes a case for drastic early cuts while assuming large damages far into the future. But Nordhaus argues that if one used a more traditional time discount rate and made typical market assumptions about returns on capital investments, Stern's result would more closely resemble the standard predictions of economic models about carbon taxes and emissions reductions. In other words, Nordhaus is claiming that Stern overstates the costs of climate change. In order to evaluate these two contending positions, readers will need to test their own feelings about the ethical impacts of various welfare assumptions regarding the relative welfare of present and future generations.

IPCC Report Lays Out Options for Taming Greenhouse Gases

JOHN BOHANNON, WITH REPORTING BY ELI KINTISCH

B ANGKOK—Reining in climate change won't bankrupt the world economy and won't require technological miracles. But we'll have to start soon. That is the mostly upbeat conclusion from Working Group III of the Intergovernmental Panel on Climate Change (IPCC), which met behind closed doors for three days last week here in the Thai capital.

The fruit of the working group's labor is a 35-page document that lays out options—and their price tags—for reducing greenhouse gas emissions to head off catastrophic climate change. The most ambitious plan, which would stabilize greenhouse gas levels in the atmosphere [measured in equivalents of carbon dioxide (CO_2)] below 535 parts per million (ppm),

---

This article first appeared in *Science* (11 May 2007: Vol. 316, no. 5826). It has been revised for this edition.

would come with an estimated 3% decrease in global gross domestic product (GDP) by 2030, compared with business as usual. Less ambitious targets come cheaper. The easiest option— aiming for under 710 ppm, 50% higher than the current atmospheric concentration of long-lived greenhouse gases of 460 ppm—could yield a small net gain for the global economy.

The report—the executive summary, written by 33 of the several hundred contributing authors, of a review of major economic modeling studies due to be released in September— concludes that getting from today's greenhouse gas-intensive economy to any of these targets is achievable with currently available tools such as shifting to alternative energy sources, boosting energy efficiency, and reducing deforestation, coupled with a suitable mix of caps, taxes, and economic incentives. But other scientists warn that reality will present harder choices than the models suggest. "The only reason for economists to make forecasts is to make astrologers

> A broad portfolio of alternative energy sources could cut projected annual CO_2 emissions in the year 2030 by five to seven gigatons (Gt) at no cost, thanks to savings from energy efficiency.

look good," says Martin Hoffert, a physicist at New York University who has criticized earlier IPCC studies.

Last-Ditch Editing

Reaching consensus on these take-home messages was easier than expected. Media reports had predicted bitter disputes between IPCC member countries. For example, China was expected to insist on softening statements that might suggest that its fast-growing and fossil-fueled economy might need to be slowed, whereas the United States was expected to bully for nuclear power. But in fact, says Dennis Tirpak, a climate policy analyst who heads the climate change unit at the Organisation for Economic Co-operation and Development in Paris and one of the summary's authors, "the atmosphere was quite civilized."

China did put its foot down—over the adjective used to characterize the scientific evidence behind estimates of the cost of achieving emissions targets. China urged that the quality be downgraded from "high" to "medium." The motivation was "only to protect the scientific integrity of the IPCC," says coauthor Dadi Zhou, a climatologist and deputy director of the Energy Research Institute in Beijing. Others who spoke

with *Science* agree. "China had a valid point, and we adopted it," says coauthor Jayant Sathaye, an energy policy analyst at Lawrence Berkeley National Laboratory in California.

In the end, only two short passages in the report fell short of unanimous approval. One was four lines stating that with a price of $50 for a ton of emitted CO_2, nuclear energy would be cost-effective in providing nearly a fifth of global electricity—with the caveat that "safety, weapons proliferation, and waste remain as constraints." Even that cautious endorsement sparked what Sathaye calls an "adrenaline-fueled" discussion ending with firmly antinuclear Austria insisting on a footnote saying that it "could not agree with this statement." The other sticking point was a passage on forestry, which drew fire on technical grounds from a delegate from Tuvalu.

The final result is a document that strikes a far more optimistic tone than did the previous three mitigation reports. At least, that was the mood of the IPCC's buoyant press release, which has been echoed by the media since its release.

Climate Crystal Ball

But hidden within the text of the report are abundant references to uncertainties and caveats that have gone largely unmentioned.

For one, many scientists are muttering, the report is only as good as its models. To explore mitigation options, the IPCC uses two distinct strategies. Bottom-up models break the economy down into sectors and predict how different mixes of technologies will cut carbon emissions in each. Top-down models simulate whole economies to compare how different global strategies, such as carbon taxes or fixed greenhouse-gas stabilization targets, will play out through market forces. Each approach has its drawbacks. Bottom-up models tend to ignore economics, whereas top-down models smooth over the differences between regions and sectors. In 2001, the two approaches were often at odds. The good news, says Sathaye, is that "for the first time, the range of results from bottom-up and top-down models are starting to converge." However, enormous wiggle room remains.

One problem is that bottom-up models don't cope well with lifestyle: the preferences that drive people to choose one mix of technologies over another. For example, the report suggests that a broad portfolio of alternative energy sources, such as solar and biofuels, could cut projected annual CO_2 emissions in the year 2030 by five to seven gigatons (Gt) at no cost at all, thanks to savings in energy efficiency. But that conclusion is misleading, says author Richard Richels, an economic modeler at the Electric Power Research Institute in Palo Alto, California, because it ignores the implicit cost of making people choose something they don't want. "If it's advantageous, why aren't people doing it?" Richels asks.

Since 2001, researchers have worked to make the models more realistic by incorporating such "market feedback," says Billy Pizer, an economist with Resources for the Future in Washington, D.C., who coauthored a related chapter in the full mitigation report. But it's one thing to account for people's illogical behavior and quite another to persuade them to change it. "It's stuff that pays for itself that people don't do," he says.

Steady progress has been made with top-down models, says Jae Edmonds of the College Park, Maryland, office of the Pacific Northwest National Laboratory. The modelers are now accounting for more regional details, such as the availability of land area for biofuels and the potential for storing coal-plant carbon emissions underground. They have also expanded the models to include emissions of greenhouse gases other than CO_2, such as methane. Doing so has lowered the top-down estimates of mitigation costs. "The reason is that you have other opportunities to reduce emissions," says Sathaye. For example, a landfill emitting methane can be cheaper to deal with than a coal plant, but such advantages were lost in previous simulations.

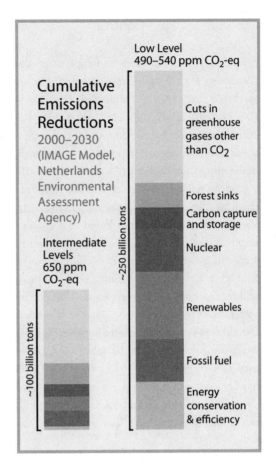

IMAGE 1 Diet plan. The IPCC report drew on models that calculated global portfolios of emissions reductions needed to reach various target levels of greenhouse gases in the atmosphere. Adapted from the IPCC Fourth Assessment Report.

But top-down models can still run aground on the shoals of international politics. One rosy prediction is that an imposed cost of $100 per ton of CO_2—equivalent to an extra $1 per gallon at the pumps—could yield a cut of 17 to 26 Gt of CO_2 by 2030, as much as 38% of estimated emissions under a fairly carbon-intensive forecast. But this assumes that the whole world participates in carbon trading and that markets are free and transparent. Given current Indian and Chinese wariness toward carbon caps, says Pizer, "that's not politically likely."

Spin Control

Now that the debate over the content of the 1000-page Fourth Assessment Report is done, the battle is shifting to its interpretation. Many IPCC scientists say they are uneasy with the optimistic spin put on the report. "I think something that is being underplayed . . . is the scale of the mitigation challenge," says Brian O'Neill, a climate policy modeler at the International Institute for Applied Systems Analysis in Vienna, Austria, who contributed to a chapter on mitigation scenarios. "To limit warming to something near the European Union's stated goal of 2°C, global emissions have to peak within the next decade or two and be cut by 50% to 80% by mid-century." That's a tall order, O'Neill says—and it could get a lot taller if global temperatures turn out to be more sensitive to increases in greenhouse gases than the IPCC has been assuming. "My point is not that there should be more gloom and doom," says O'Neill, but "a message that says that we have to stay below 2°C, but don't worry, it will be easy and cheap, just doesn't add up."

Other researchers say the report's insistence that current mitigation strategies can suffice gives short shrift to future research. That's a mistake, says Hoffert: "It is ludicrous to think a greenhouse gas emissions price, cap, or tax alone will get you to stable concentrations of [greenhouse gases]." New technologies will be critical, he says, and unless policymakers pave the way with measures such as a gradually increasing carbon tax, they will not be competitive. And Richels fears that if the takeaway message is that mitigation is cheap, societies "may not be as motivated to invest in the future" for such research.

Overall, the question of whether mitigation is "affordable"—be it 0.3% or 3% of global GDP—is "a difficult one to answer," says Sathaye. But some say that when stakes are overwhelmingly high, purely economic reasoning misses the boat. "What did World War II cost us economically?""asks Hoffert. "Does the question even make sense?"

The Limits of Consensus

MICHAEL OPPENHEIMER, BRIAN C. O'NEILL, MORT WEBSTER,
SHARDUL AGRAWALA

The Intergovernmental Panel on Climate Change (IPCC) has just delivered its Fourth Assessment Report (AR4) since 1990. The IPCC was a bold innovation when it was established, and its accomplishments are singular (1, 2). It was the conclusion in the IPCC First Assessment Report that the world is likely to see "a rate of increase of global mean temperature during the next century . . . that is greater than seen over the past 10,000 years" (3) that proved influential in catalyzing the negotiation of the United Nations Framework Convention on Climate Change. The conclusions of the Second Assessment with regard to the human influence on climate (4) marked a paradigm shift in the policy debate that contributed to the negotiation of the Kyoto Protocol. IPCC con-

clusions from the Third, and now the Fourth, Assessments have further solidified consensus behind the role of humans in changing the Earth's climate.

The emphasis on consensus in IPCC reports, however, has put the spotlight on expected outcomes, which then become anchored via numerical estimates in the minds of policymakers. With the general credibility of the science of climate change established, it is now equally important that policymakers understand the more extreme possibilities that consensus may exclude or downplay (5).

For example, the Working Group I "Summary for Policymakers" (SPM) of AR4 anticipates a rise in sea level of between 18 and 59 cm by the year 2100 (6), a "model-based range" composed largely of thermal expansion of oceans, melting of nonpolar glaciers, and the gradual response of ice sheets. The range does not include the potential for increasing contributions from rapid dynamic processes in the Greenland Ice

---

This article first appeared in *Science* (14 September 2007: Vol. 317, no. 5844). It has been revised for this edition.

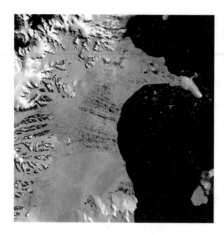

IMAGE 1 Not captured by ice sheet models. (Left) The Larsen B ice shelf along the Antarctic Peninsula on 31 January 2002. (Right) A large section has disintegrated, 5 March 2002. Glaciers behind the collapsed section of the ice shelf subsequently accelerated their discharge into the ocean, apparently because of the loss of buttressing by the ice shelf. Neither rapid collapse nor buttressing are captured by ice sheet models, and both could substantially affect the rate of future sea level rise as larger ice shelves to the south in West Antarctica warm (26).
Source: National Snow and Ice Data Center, Boulder, CO.

Sheet and West Antarctic Ice Sheet (WAIS), which have already had a significant effect on sea level over the past 15 years and could eventually raise sea level by many meters. Lacking such processes, models cannot fully explain observations of recent sea level rise, and accordingly, projections based on such models may seriously understate potential future increases. Although the AR4 SPM recognizes the possibility of a larger ice sheet contribution, its main quantitative results indicate the opposite: uncertainty in sea level rise is smaller, and its upper bound is lower, for the 21st century than was indicated in the Third Assessment Report (7). On the related question of sea level rise beyond the 21st century, whereas the Third Assessment's SPM provided a numerical estimate of a potential contribution from WAIS, the AR4 Working Group I SPM doesn't mention WAIS at all. This omission presumably reflects a lack of consensus arising from the inadequacy of ice sheet models for WAIS made so apparent by recent observations.

Nevertheless, alternatives to model-based approaches, such as empirical analysis and expert elicitation, were available for exploring uncertainty in 21st-century (8) and long-term sea level rise (9), respectively. Such information certainly would have been useful to policymakers, particularly for WAIS, which contains enough ice to raise sea level by about 5 m.

Setting aside or minimizing the importance of key structural uncertainties in underlying processes is a frequent outcome of the drive for consensus (5, 10). For example, ranges of projected warming and atmospheric composition in AR4 include an amplifying effect of interactions between climate and the carbon cycle. However, the estimated uncertainty in this effect is based largely on models that omit a number of poorly understood processes (11), such as feedbacks on carbon contained in permafrost; changes in marine ecosystem structure; and responses to land-use history, nutrient limitation, and air pollution effects. These models also share similar assumptions about the temperature sensitivity of carbon fluxes from soils based on experimental results that cannot be reliably scaled to the ecosystem level (12). A fuller accounting of uncertainty would be more appropriate.

Similarly, the narrowing of uncertainty (relative to previous assessments) associated with potential changes in the **meridional overturning circulation** relies on agreement across models, but the structural uncertainty in all the models means that less may be known than suggested by the numerical estimates (13).

Like models of physical processes, conclusions drawn on the basis of socioeconomic models may also be subject to premature consensus. Estimates of the costs of mitigating emissions come primarily from models that omit endogenous technical change, a poorly understood process. This omission could cause a significant bias, not only in mitigation costs, but also in the stringency of near-term mitigation that may be justified for a given damage function or stabilization target (14–16). Similarly, the conventional use of the range of emissions described by the IPCC Special Report on Emissions Scenarios (SRES) marker scenarios as a key determinant of uncertainty in projecting climate change, sea level rise, impacts, and mitigation costs may be misguided. The SRES scenarios were intended to be representative of scenarios available in the literature at the time they were produced, with no explicit goal of spanning the full range of uncertainty. The SRES assessment made no attempt to judge whether emissions pathways outside the range it covers could plausibly occur. In fact, pathways outside that range were known at the time, and more have been developed since the publication of SRES (17).

To be sure, the underlying IPCC chapters do detail the limitations and uncertainties associated with such conclusions. But the caveats are often cryptic or lost entirely in the highly influential SPMs. This inevitably leads to an anchoring by both policymakers and scientists around any numerical estimates that are reported in these summaries.

Ignoring the implications of structural uncertainty in models of key aspects of the climate system is reminiscent of the way assessments treated the uncertainty in ozone photochemical models. Projections of ozone depletion were

KEY TERM

Meridional overturning circulation: The MOC is a global-scale northward ocean circulation that carries warm, salty Atlantic Ocean water from the tropics to near Greenland, where the water cools and sinks to the abyss, circulates worldwide, and then resurfaces near Antarctica. During glacial times, abrupt shifts in global climate were associated with periods when this circulation shut off, and there is some concern that the buildup of greenhouse gases may bring about similar abrupt climate changes in the future.

made from 1974 onward based on improved understanding of gas-phase chemistry (18). Knowledge of stratospheric chemistry was then transformed by the report in 1985 of large, seasonal Antarctic depletion (the "ozone hole"); the validation in 1987 of its origin in halogen photochemistry; and subsequent identification of depletion at the mid-latitudes and in the Arctic (19, 20). Various heterogeneous chemical reactions, discounted by most researchers years before and absent from nearly all model simulations (21), were shown to be the missing photochemical processes required to explain observed depletion. Their potential implications were of concern to some scientists (22), but this structural uncertainty was generally downplayed in assessments until the ozone hole was reported.

Avoiding Premature Consensus

The IPCC has made progress over four assessment cycles in its treatment of uncertainties.

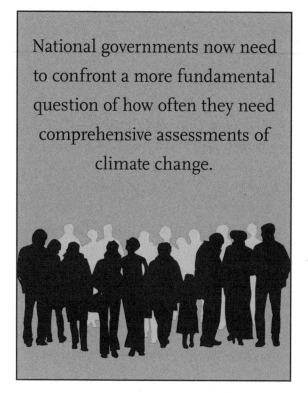

National governments now need to confront a more fundamental question of how often they need comprehensive assessments of climate change.

However, this progress is limited and uneven across its Working Groups. Several additional modifications to the current practice could reduce the risk of ignoring or underemphasizing critical uncertainties.

First, given the anchoring that inevitably occurs around numerical values, the basis for quantitative uncertainty estimates provided must be broadened to give observational, paleoclimatic, or theoretical evidence of poorly understood phenomena comparable weight with evidence from numerical modeling. In areas in which modeling evidence is sparse or lacking, IPCC sometimes provides no uncertainty estimate at all. In other areas, models are used that have quantitatively similar structures, leading to artificially high confidence in projections (e.g., in the sea level, ocean circulation, and carbon cycle examples above). One possible improvement would be for the IPCC to fully include judgments from expert elicitations (23), as Working Group II has sometimes done. Beyond this, increased transparency,

including a thorough narrative report on the range of views expressed by panel members, emphasizing areas of disagreement that arose during the assessment, would provide a more robust evaluation of risk (24). It would be critical to include this information not only in the chapters, but in the summaries for policymakers as well.

Second, IPCC should revise its procedure for expert review to guard against overconfidence. External reviewers should ferret out differences between chapters or author subgroups, and a special team of authors could be instructed to examine the treatment of unlikely but plausible processes, perhaps in a separate chapter. Integration of risk assessment across Working Groups in advance of drafting of the Synthesis Report would highlight internal discussions and disagreements. At the end of an assessment cycle, a small external team of ombudsmen should review key problematic issues (of a scientific nature) that may have emerged from the report and should recommend modifications of approaches for handling these areas in subsequent reports.

Third, IPCC could also formalize a process of continuous review of its structure and procedures. A useful example is provided by the history of IPCC emissions scenario development, which included a series of reviews for production of the SA90, IS92, and SRES scenarios (25).

Fourth, and perhaps most important, national governments now need to confront a more fundamental question of how often they need comprehensive assessments of climate change. Addressing the special risks entailed in particular aspects of the climate system, like the ice sheets or carbon cycle, might be better approached by increasing the number of concise, highly focused special reports that can be completed relatively quickly by smaller groups, perhaps even by competing teams of experts. At this juncture, full assessments emphasizing consensus, which are a major drain on participants and a deflection from research, may not be needed more than once per decade.

References and Notes

1. S. Agrawala, *Clim. Change* 39, 621 (1998).

2. N. Oreskes, *Science* 306, 1686 (2004).

3. IPCC, Summary for Policymakers, in *Scientific Assessment of Climate Change: Report of Working Group I*, J. T. Houghton *et al.*, Eds. (Cambridge Univ. Press, Cambridge, 1990).

4. IPCC, Summary for Policymakers, in *Climate Change 1995: The Science of Climate Change*, J. T. Houghton *et al.*, Eds. (Cambridge Univ. Press, Cambridge, 1995).

5. G. Patt, *Risk Decis. Policy* 4, 1 (1999).

6. IPCC, Summary for Policymakers, in *Climate Change 2007: The Physical Science Basis: Contribution of Working Group I to the Fourth Assessment Report of the Intergovernmental Panel on Climate Change*, S. Solomon *et al.*, Eds. (Cambridge Univ. Press, New York, 2007).

7. IPCC, Summary for Policymakers, in *Climate Change 2001: The Physical Science Basis*, J. T. Houghton *et al.*, Eds. (Cambridge Univ. Press, Cambridge, 2001).

8. S. Rahmstorf, *Science* 315, 368 (2007).

9. D. G. Vaughan, J. R. Spouge, *Clim. Change* 52, 65 (2002).

10. M. Oppenheimer, B. C. O'Neill, M. Webster, paper presented at the Conference on Learning and Climate Change, International Institute for Applied Systems Analysis, Laxenburg, Austria, 10 April 2006.

11. P. Friedlingstein *et al.*, *J. Clim.* 19, 3337 (2006).

12. J. M. Melilo *et al.*, *Science* 298, 2173 (2002).

13. K. Zickfeld *et al.*, *Clim. Change* 82, 235 (2007).

14. L. H. Goulder, S. H. Schneider, *Resour. Energy Econ.* 21, 211 (1999).

15. E. Baker, *Resour. Energy Econ.* 27, 19 (2005).

16. R. Gerlagh, B. van der Zwaan, *Resour. Energy Econ.* 25, 35 (2003).

17. M. Webster *et al.*, *Atmos. Environ.* 36, 3659 (2002).

18. E. Parson, *Protecting the Ozone Layer: Science and Strategy* (Oxford Univ. Press, New York, 2003).

19. World Meteorological Organization, *Report of the International Ozone Trends Panel—1988* (Global Ozone Research and Monitoring Project, World Meteorological Organization, Geneva, Report 18, 1988).

20. World Meteorological Organization, *Scientific Assessment of Ozone Depletion—1991* (Global Ozone Research and Monitoring Project, World Meteorological Organization, Geneva, Report 25, 1991).

21. World Meteorological Organization, *Atmospheric Ozone 1985* (Global Ozone Research and Monitoring Project, World Meteorological Organization, Geneva, Report 16, 1986).

22. F. S. Rowland, *Am. Sci.* 77, 36 (1989).

23. G. Morgan, M. Henrion, *Uncertainty: A Guide to Dealing with Uncertainty in Quantitative Risk and Policy Analysis* (Cambridge Univ. Press, Cambridge, 1990).

24. A. Patt, *Glob. Environ. Change* 17, 37 (2007).

25. The SA90 scenarios were published in the IPCC First Assessment Report. The IS92 scenarios were published in the 1992 Supplementary Report to the IPCC Assessment.

26. T. Scambos, J. Bohlander, B. Raup, compilers, Images of Antarctic ice shelves [2001, updated 2002], National Snow and Ice Data Center, Boulder, CO.

Commentary:
A Closer Look at the IPCC Report

SUSAN SOLOMON, RICHARD ALLEY, JONATHAN GREGORY,

PETER LEMKE, MARTIN MANNING

In their Policy Forum ("The Limits of Consensus," 14 September 2007, p. 1505), M. Oppenheimer *et al.* make several misleading statements. They suggest that a premature drive for consensus led Working Group I to understate the risk of large future sea level rise in the Intergovernmental Panel on Climate Change (IPCC) Fourth Assessment Report (AR4). They assert that the "Summary for Policymakers" (SPM) of the AR4 did not properly consider increasing contributions from rapid dynamic changes in the ice sheets of Greenland and West Antarctica (WAIS). However, in quoting the SPM discussion of sea level rise, they ignore its explicit statements on the subject, such as "dynamical processes related to ice flow not included in current models but suggested by

recent observations could increase the vulnerability of the ice sheets to warming, increasing future sea level rise"; the model projections "[do not] include the full effect of ice sheet flow because a basis in published literature is lacking"; and, crucially, "larger values cannot be excluded, but understanding of these effects is too limited to assess their likelihood or provide a best estimate or an upper bound for sea level rise" (1).

We agree with Oppenheimer *et al.* that paleoclimatic observations should be considered in assessing possible long-term future sea level rise and polar ice sheet changes, but we dispute their inference that the SPM omitted the available information. The SPM explicitly noted that "global average sea level in the last interglacial period (about 125,000 years ago) was likely 4 to 6 m higher than during the 20th century, mainly due to the retreat of polar ice" from Greenland and possibly Antarctica as well. The SPM refers to the whole of Antarctica because

This article first appeared in *Science* (25 January 2008: Vol. 319, no. 5862). It has been revised for this edition.

of the possibility of differing behavior for the East Antarctic Ice Sheet (for which there is currently some evidence for mass gain, as opposed to mass loss of WAIS), in order to communicate with policymakers whose interest lies in understanding the total contribution to sea level rise.

Oppenheimer *et al.* offer a number of suggestions for handling uncertainty, but they do not address the fact that quantitative model projections of ice sheet dynamical changes cannot yet be made because of the inadequacy of current scientific knowledge. Observations do not presently offer a clear way to progress past these model limitations, in part because of discrepancies among published studies: whereas some suggest that currently observed flows in outlet glaciers may be transient and thus have limited implications for long-term sea level rise, others suggest the opposite.

IPCC assesses the literature; it does not conduct new research. In our view, providing numerical estimates of potential sea level rise due to processes not yet quantified in the literature (whether by expert elicitation, as suggested by Oppenheimer *et al.*, or by another process) would lead to inappropriate "anchoring around numerical values" of exactly the type that Oppenheimer *et al.* warn against. Far from minimizing structural uncertainties, or driving for a "premature consensus" as Oppenheimer *et al.* suggest, the SPM text of the AR4 appropriately does the exact opposite by explicitly stating that "understanding of these processes is limited and there is no consensus on their magnitude."

References and Notes

1. IPCC, Summary for Policymakers (SPM), in *Climate Change 2007: The Physical Science Basis: Contribution of Working Group I to the Fourth Assessment Report of the Intergovernmental Panel on Climate Change*, S. Solomon *et al.*, Eds. (Cambridge Univ. Press, New York, 2007).

Response to Commentary:
A Closer Look at the IPCC Report

MICHAEL OPPENHEIMER, BRIAN O'NEILL, MORT WEBSTER,
SHARDUL AGRAWALA

W e disagree with Solomon *et al.* that our Policy Forum was misleading. We correctly noted that model-based numerical ranges for 21st-century sea level rise presented in the Working Group I (WGI) "Summary for Policymakers" (SPM) (Table SPM-3) did not account for the uncertainty resulting from potential increases in the rapid dynamic response of ice sheets. Solomon *et al.* challenge this assertion by pointing instead to qualitative statements in the SPM, implying that the latter provide a satisfactory accounting of uncertainty. But the distinction between numerical values highlighted in a prominent table and narrative qualifications of such numbers is critically important. Numbers are powerful, grabbing the readers' attention, whereas

qualifications are often ignored. For example, the tabular values, indicating a maximum sea level rise of 59 cm during the 21st century, are cited frequently in the public discussion absent any qualification.

We did not imply, as Solomon *et al.* argue, that the WGI SPM omitted information from paleoclimate studies in evaluating uncertainty in sea level rise beyond the 21st century. We suggested that it gave too much credence to ice sheet models compared with other sources of information. For example, in reporting only a model-based estimate for the time scale of a long-term contribution (from Greenland), the WGI SPM gives short shrift to the implications of observations of fast responses in the ice sheets in both Greenland and western Antarctica, narrative qualifications to the contrary notwithstanding. Such an approach understates the range of opinion in the relevant expert community on the potential magnitude and rate of the ice sheet contribution as indicated by studies

---

This article first appeared in *Science* (25 January 2008: Vol. 319, no. 5862). It has been revised or this edition.

reviewed during the Fourth Assessment Report (AR4) (1). Further perspective on this question is provided by the AR4 Synthesis Report (2).

Finally, contrary to Solomon *et al.*'s assertion, our suggestions for improving the treatment of uncertainty were made specifically with the shortcomings of ice sheet modeling in mind. It makes little sense to highlight model-based projections of sea level rise when models that are supposed to account for the ice sheet component have failed the test against reality. Other approaches provide important additional perspectives. For example, the fact that two independent semi-empirical analyses estimating uncertainties in future sea level rise have been published recently (3, 4) suggests that observation-based methods yield important insights where models are deficient.

We do not propose that IPCC "conduct new research." Rather, we argue that it take full advantage of what has already been produced. IPCC also has the flexibility to fill gaps in modeling and analysis where the completeness of assessment calls for it, and it has done so many times. In anticipation of a Fifth Assessment, and realizing that ice sheet models may not improve rapidly, IPCC should encourage the development of a more comprehensive approach to uncertainty. As it has done for other arenas, such as emissions scenarios or abrupt climate change, IPCC could spur research into empirical approaches, formalized expert elicitation, and comprehensive analysis of paleo-ice extent and sea level, each carried out with a specific view toward informing quantitative judgments on the range of future sea level. Holding workshops on this problem over the next few years would fit neatly into IPCC tradition.

Three of us are authors of AR4, well aware of the difficulty of assessment. A premise of our Policy Forum is that IPCC has done a superb job of establishing the scientific consensus. But in a high-stakes problem like global warming, governments need to calibrate policy to the full range of plausible outcomes, for sea level rise and for all other key uncertainties.

References

1. G. A. Meehl *et al.*, Global climate projections, in *Climate Change 2007: The Physical Science Basis, Contribution of Working Group I to the Fourth Assessment Report of the Intergovernmental Panel on Climate Change*, S. Solomon *et al.*, Eds. (Cambridge Univ. Press, Cambridge, 2007).

2. IPCC, *Fourth Assessment Report, Climate Change 2007: Synthesis Report*; www.ipcc.ch/ipccreports/ar4-syr.htm.

3. M. F. Meier *et al.*, *Science* 317, 1064 (2007).

4. S. Rahmstorf, *Science* 315, 368 (2007).

Simulating Arctic Climate Warmth and Ice Field Retreat in the Last Interglaciation

BETTE L. OTTO-BLIESNER, SHAWN J. MARSHALL,
JONATHAN T. OVERPECK, GIFFORD H. MILLER, AIXUE HU,
CAPE LAST INTERGLACIAL PROJECT MEMBERS

In the future, Arctic warming and the melting of polar glaciers will be considerable, but the magnitude of both is uncertain. We used a global climate model, a dynamic ice sheet model, and paleoclimatic data to evaluate Northern Hemisphere high-latitude warming and its impact on Arctic ice fields during the Last Interglaciation. Our simulated climate matches paleoclimatic observations of past warming, and the combination of physically based climate and ice sheet modeling with ice core constraints indicate that the Greenland Ice Sheet (GIS) and other circum-Arctic ice fields likely contributed 2.2 to 3.4 m of sea level rise during the Last Interglaciation.

This article first appeared in *Science* (24 March 2006: Vol. 311, no. 5768). It has been revised for this edition.

Determining the sensitivity of the Arctic climate system to anomalous forcing and understanding how well climate models can simulate the future state of the Arctic are critical priorities. Over the past 30 years, Arctic surface temperatures have increased 0.5°C per decade [1]; September Arctic sea-ice extent has decreased 7.7% per decade [2]; and the seasonal ablation area for Greenland has increased, on average, by 16% [3–5]. The global climate models being used to estimate future scenarios of Arctic warmth give polar warming of 0.7°C to 4.4°C—a large range—as well as a reduction of Arctic sea ice of up to 65% at the time of the doubling of atmospheric CO_2 [6]. The Last Interglaciation (LIG, about 130,000 to 116,000 years ago) is the last time that the Arctic experienced summer temperatures markedly warmer than those in the 20th century and the late Holocene, and it also featured a significantly reduced GIS. Climate models need to be able to reproduce this large, warm climate change in the Arctic if they are to be trusted in their representation of Arctic processes and their predictions for the future.

Paleorecords indicate much warmer Arctic summers during the LIG. Storm beaches and ancient barrier islands with mollusks of LIG age indicate that the open water north of Alaska was more extensive and lasted seasonally longer (7). Boreal forest communities expanded poleward by as much as 600 to 1000 km in Russia (8), reaching the coast everywhere except in Alaska (9) and central Canada. Total gas evidence from LIG ice in the Greenland Ice Core Project (GRIP) ice core indicates that the Summit region remained ice covered, although possibly up to about 500 m lower than the ice level at present, at some time in the LIG (10). In contrast, basal ice at Dye-3 (southern Greenland); in the Agassiz, Devon, and Meighen ice caps in the Canadian Arctic; and possibly in Camp Century (northwest Greenland) suggest that these drill sites were ice-free during the LIG (10, 11). The increased presence of vegetation over southern Greenland is reconstructed from plant macrofossils (12) and fern spores (13). Elsewhere, pollen, insects, marine plankton, and other proxies document the magnitude of LIG summer warmth across the Arctic (14).

We conducted climate simulations for the LIG with a global coupled ocean–atmosphere–land–sea-ice general circulation model [National Center for Atmospheric Research (NCAR) Community Climate System Model (CCSM)] (15). We also used ice sheet simulations with a three-dimensional, coupled ice-and-heat-flow model (16), which spans the entire western

Arctic from 57°N to 85°N, including Greenland and smaller-scale ice caps in Iceland and the Canadian Arctic (14). We chose to simulate the start of the LIG, approximately 130,000 years ago (130 ka), reflecting evidence of early LIG summer Arctic warmth (14) and of an LIG sea level high stand of 4 to 6 m above present day likely by 129 ka ± 1000 years (17, 18).

Our climate simulation was forced with the large insolation anomalies of the LIG at 130 ka—anomalies driven by changes in Earth's orbit, which are known to have caused warm Northern Hemisphere climate (19–21). The anomalous forcing for the start of the LIG (130 to 127 ka) was concentrated in the late spring and early summer because of the nature of Milankovitch solar anomalies. This timing is important because the GIS is extremely sensitive to warm, early-summer conditions (22). Positive solar anomalies at the top of the atmosphere exceeded 60 watts per square meter (W/m^2) at high northern latitudes in May to June for 130 ka (Fig. 1) and until 127 ka. After 127 ka, these positive solar anomalies decreased considerably, and by 125 ka, they were less than the maximum May-to-June solar anomalies of about 45 W/m^2 during the Holocene, which were not enough to melt the GIS much beyond its modern configuration. Annual mean changes of insolation at high northern latitudes for 130 ka were much

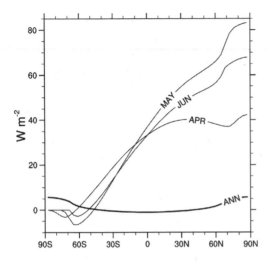

FIGURE 1. Anomalies of solar radiation incoming at the top of the atmosphere at 130 ka relative to present as a function of latitude. Large positive anomalies occur over the Arctic during boreal late spring and early summer. Northern high-latitude annual mean anomalies are small and positive, whereas global annual mean anomalies are close to zero. Similar high-latitude anomalies occur for boreal late spring and early summer through 127 ka, diminishing considerably after that time. ANN=annual.

smaller, less than 6 W/m^2; global annual mean forcing was very small, 0.2 W/m^2.

A comparison between proxy-based reconstructions for the Arctic LIG and those simulated with the CCSM climate model shows good agreement (Fig. 2). Solar anomalies drive significant simulated summer (June, July, and August) warming in the Arctic (for 60°N to 90°N, an average warming of 2.4°C with significant regional variation) (Fig. 2). In agreement with paleoclimatic observations, simulated warming in excess of 4°C occurs in the northern Hudson Bay–Baffin Island–Labrador Sea region and across to the seas adjacent to northern and eastern Greenland. Greenland warms by 3°C or greater along the edges of the ice cap and by 2.8°C in central Greenland in CCSM, somewhat less than observed. Less summer warming is indicated for northern Europe (2°C to 3°C), as well as Alaska and western Canada (0°C to 2°C), in both the proxy reconstruction and CCSM. The simulated warming over Siberia is 2°C to 4°C, somewhat less than the paleodata in parts of this region.

The insolation anomalies result in increased sea-ice melting early in the summer season, with reduced sea ice extending from April into November. The minimum LIG Arctic sea-ice area simulated in August and September is 50% less than that simulated for present

day, with 50% summer coverage in the Arctic Ocean only occurring poleward of 80°N (see fig. S1 online). The reduction of summer sea ice leads to simulated warming of the Arctic Ocean north of Alaska by about 2°C. The North Atlantic warms by 1°C to 2°C, with above-freezing temperatures extending northward along circumcoastal regions, and substantial warming extending as far east as Severnaya Zemlya in CCSM. The CCSM is in accord with marine records that indicate modest warming (0°C to 4°C) at LIG (Fig. 2).

The simulated temperature response of the Arctic to altered LIG insolation is significant over the ice sheet and the surrounding waters of Greenland. Feedbacks associated with reduced sea ice and a warmer North Atlantic Ocean and Labrador Sea act to warm all of Greenland during the summer months, with the southern quarter and far northern coastal regions averaging above freezing for multiple months. Maximum daily surface temperatures during the summer are above freezing over the entire ice sheet. Annual snow depths decrease significantly along the southern, western, and northern edges of the ice sheet (17). These decreases are primarily due to melting with warmer surface temperatures. Modeled precipitation rates are generally not significantly different from present, except for marginally significant increases in northwest

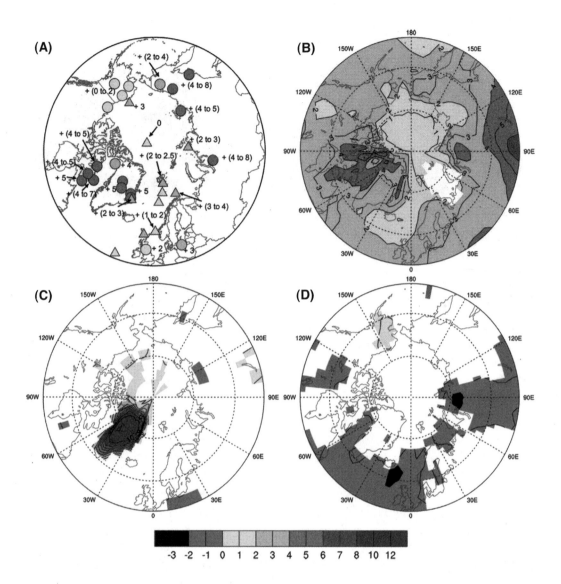

FIGURE 2. Arctic summer surface temperature anomalies. (A) Maximum observed LIG summer temperature anomalies relative to present derived from quantitative [terrestrial (circles) and marine (triangles)] paleotemperature proxies as part of CAPE Last Interglacial Project (14). (B) LIG summer (June, July, and August) temperature anomalies relative to present simulated by CCSM for 130 ka. (C) Additional LIG summer warming for our no–Greenland Ice Sheet sensitivity simulation relative to our LIG simulation with the GIS. (D) Summer temperature anomalies for a freshwater anomaly of 0.1 sverdrup to the North Atlantic for 100 years simulated by CCSM. For the CCSM simulations, only anomalies significantly different from natural model variability at 95% confidence interval are shown.

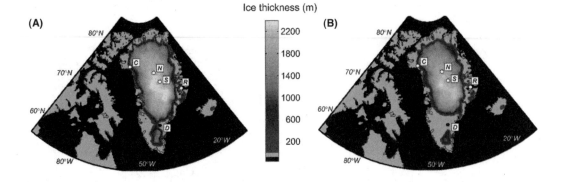

FIGURE 3. Simulated ice sheet thickness maps for LIG climate forcing. (A) Configuration for when the location of Dye-3 ice core becomes ice-free. This configuration gives the minimum sea level rise (2.2 m) that the Arctic likely contributed during the LIG. (B) Configuration for a paleodivide elevation of 570 m lower than present. This configuration gives the maximum Arctic sea level rise (3.4 m) that the Arctic likely contributed during the LIG. Greenland ice core observations indicate early LIG ice (white circles) at Renland (marked R; 71.3°N, 26.7°W), North Greenland Ice Core Project (NGRIP) (marked N; 75.1°N, 42.3°W), Summit [GRIP and Greenland Ice Sheet Project 2 (GISP2)] (marked S; 72.5°N, 37.3°W), and possibly Camp Century (marked C; 77.2°N, 61.1°W), but not (red circle) at the Dye-3 ice core site (marked D; 65.2°N, 43.8°W).

and central Greenland and southeast Iceland, associated with warmer nearby oceans.

Simulated margins of the GIS, as well as smaller Arctic ice fields, respond immediately to the warmer spring and summer temperatures. The total ice field ablation area increases from 2.64×10^5 km^2 for modern time to 5.25×10^5 km^2 in the LIG, with melt rates near the ice margins increasing by as much as 1 m/year. Greater snow accumulation compensates for elevated melt rates in southeast Iceland, central Greenland, and isolated coastal sites in Greenland, giving small, positive mass-balance perturbations for the LIG of about 0.2 m/year. Elsewhere in the western Arctic, increased melt rates and an extended ablation season lead to negative mass-balance perturbations, with initial losses of −0.3 to −0.6 m/year in most of coastal Greenland. On average over the western Arctic, the mass-balance perturbation is −0.19 m/year.

As the simulated ice caps retreat over several millennia in response to the orbitally induced warming, loss of ice mass leads to surface

lowering, an amplification of the mass-balance perturbation, and accelerated retreat. The sustained negative mass-balance perturbations cause the almost complete demise of ice fields in the Queen Elizabeth Islands, consistent with the ice core inferences discussed in Koerner (11, 23). In approximately 2000 years, the GIS has retreated such that Dye-3 becomes ice-free, in agreement with LIG paleorecords (Fig. 3A) (11–13). In this configuration, Greenland and the western Arctic ice fields contribute 2.2 m of sea level rise; this is the minimum sea level rise that the Arctic likely contributed during the LIG. After an additional millennium, the simulated LIG surface drawdown at the paleodivide is about 570 m, near the constraint provided by ice core data (10). This minimal GIS configuration (Fig. 3B) yields a maximum Arctic sea level contribution of 3.4 m.

Our simulated GIS is not in equilibrium, and continued warmth would drive a smaller and lower ice sheet and continued sea level rise. However, the GIS configurations in Figure 3 are consistent, in terms of height and area,

*Science*NOW Daily News, 18 December 2007

LESSONS FROM AN INTERGLACIAL PAST
Phil Berardelli

When the Nobel Prize–winning Intergovernmental Panel on Climate Change (IPCC) issued its latest report on global warming last summer, one of its most dramatic predictions was that sea levels would rise as much as 0.58 m during the next century—enough to threaten coastal cities in Southeast Asia and North America. That's nothing, however, compared with what happened about 124,000 years ago. At a certain point during the warm "interglacial" between the last two ice ages, scientists have calculated, sea levels rose almost three times as fast. Given that the IPCC report predicts surface temperatures will reach similar levels during the next 100 years, the panel's dire forecast may not be dire enough.

Locked in ice or flowing freely, the world's amount of water remains relatively constant. But sea levels can undulate 100 m up or down over several millennia, depending on the ratio of water to ice. That's about how much seas dropped, for example, during the last ice age, which ended about 15,000 years ago—enough to allow the ancestors of Native Americans to walk from Siberia to Alaska courtesy of a land bridge that surfaced across the Bering Strait.

Now an international research team has discovered that during the warm period following the next-to-last ice age, when global temperatures reached at least 2°C above the current average, the seas rose by as much as 6 m over just a few hundred years. The team reached this conclusion by analyzing microfossil-containing sediments from the floor of the Red Sea. Those sediments preserve the ratio of certain oxygen isotopes that provide strong signals about the strength of currents and other factors. By tracking the ratio over the time the sediments span, the researchers have been able to compute the rate of flow into the Red Sea from the Indian Ocean, which indicates the level of the water. Their analysis, reported online in *Nature Geoscience*, shows that as global warming was in the process of melting the continental glaciers about 124,000 years ago, sea levels began rising at the relatively blistering rate of about 1.6 m per century.

The rapid rate in ancient times remains relevant, says geoscientist and lead author Eelco Rohling of the University of Southampton's National Oceanography Centre in the United Kingdom. It "offers a warning" to climatologists that sea level changes can depend strongly on factors that influence ice formation and melting—factors that he says current IPCC climate models understate. The per-century rate of 0.6 m that the models are predicting for the near future is 1.0 m less than the findings of Rohling's team. That difference, he says, "clearly identifies" the need to improve the climate models to reflect the impact of glaciation and melting.

Other experts aren't so sure. Geologist William Thompson of Woods Hole Oceanographic Institution in Massachusetts says that although the IPCC estimates indeed "may be too conservative" and the Red Sea research provides an "important contribution to our understanding of past sea level changes," there are "significant uncertainties" in the method used by the team, and other studies haven't shown "such high rates of sea level change."

with observed ice core data, thus providing a constraint on the Canadian Arctic and Iceland ice fields and of the GIS at their smallest size during the LIG (Fig. 3). Moreover, our use of observations and model together indicate that the minimal LIG GIS was a steep-sided ice sheet in central and northern Greenland (Fig. 3) and that this ice sheet, combined with the change in other Arctic ice fields, likely generated 2.2 to 3.4 m of early LIG sea level rise (1.9 to 3.0 m from Greenland and 0.3 to 0.4 m from Arctic Canada and Iceland). Previous modeling studies of the LIG, based on independent, ice core–derived temperature histories (24, 25) from GRIP (Greenland) $\delta^{18}O$ and Vostok (Antarctica) δD data, also suggest a minimum GIS within a few millennia of the maximum LIG insolation anomaly.

Our climate model captures the terrestrial warming within the uncertainties of proxy estimates, except over Siberia. In Siberia, the model underestimates the observed warming by up to 1°C to 2°C. Previous modeling studies have shown that Arctic vegetation changes act as a positive feedback for past periods of enhanced summer solar radiation, including the Holocene (26, 27) and the LIG (28). Our simulation lacked this positive feedback, just as this feedback is missing in most projections of future warming. However, our results indicate that CCSM does a good job of simulating much of the observed Arctic response to altered radiative forcing without this feedback, confirming that the size of the missing vegetation feedback is likely smaller than was previously estimated (14, 19, 27).

A sensitivity simulation with CCSM with all ice removed from Greenland and replaced with lower-albedo bedrock gives a measure of the large positive feedbacks associated with the large reduction of surface albedos due to ice sheet retreat and vegetation growth over Greenland (Fig. 2C). The additional warming is primarily a local response over Greenland. Surface temperatures warm an additional 7°C at Renland and Dye-3 and more than 10°C at GRIP. This warming is clearly an overestimate compared

with ice core and other proxy estimates (Fig. 2), confirming the likelihood that the GIS retreated only to the smaller and steeper configuration that we simulated (Fig. 3).

A sea level rise of 3.4 m over 3000 years is equivalent to freshwater forcing of the North Atlantic and Labrador Sea of 0.013 sverdrup. It is reasonable to assume, however, that the rate of meltwater discharge would not have been constant in time, and it could have been greater at some times than others during the drawdown of the GIS. We thus performed a sensitivity simulation with 0.1 sverdrup of water inserted into the present-day North Atlantic over 100 years. This freshwater forcing yields a simulated 25% slowdown of the North Atlantic thermohaline circulation and annual cooling of 1.5°C south of Greenland, within the range of sensitivities of 12 climate model simulations of −3.9°C to +0.7°C (29). Simulated summer cooling is generally 1°C to 2°C over much of the North Atlantic and Labrador Sea (Fig. 2D). Even with this more extreme freshwater forcing than is implied by our simulated average LIG meltwater rate, summer surface temperature anomalies over Greenland remain positive, which is an indication of the implied primacy of the large orbital forcing anomalies. These results therefore indicate that the likely impact on the North Atlantic of any Greenland meltwater would not have inhibited the meltback of Greenland.

Our results confirm that the NCAR climate model (with doubled atmospheric carbon dioxide equilibrium sensitivity of 2.2°C) captures key aspects of Arctic sensitivity to anomalous LIG forcing. Simulated summer Arctic warming is up to 5°C (as compared with a simulated global cooling of 0.7°C and Northern Hemisphere warming of 1.3°C), and simulated sea ice retreats by 50%. The climate model anomalies drive large-scale ice sheet retreat in the Western Arctic that is consistent with available ice core records. Within a few millennia, most of the ice fields in Arctic Canada and Iceland melted away, and the GIS was reduced to a steep ice dome in central and northern Greenland. We cannot comment on exactly when this ice configuration could have occurred during the LIG; there are no paleoclimatic observational constraints on the time evolution of ice sheet retreat, and the lack of meltwater-driven, ice-dynamic processes in current ice sheet models (30) prevents an evaluation of ice sheet model sensitivity. However, our results give a likely Arctic (including Greenland) contribution to LIG sea level rise above modern day of no more than 3.4 m. Despite the different mechanisms of warming, these results indicate that the impact on Arctic environments over the next century can be expected to be substantial if predicted future climate change comes to pass (17, 31).

References and Notes

1. J. C. Comiso, *J. Clim.* 16, 3498 (2003).
2. J. C. Stroeve *et al.*, *Geophys. Res. Lett.* 32, L04501 (2005).
3. W. Abdalati, K. Steffen, *J. Geophys. Res.* 106, 33983 (2001).
4. W. Krabill *et al.*, *Science* 289, 428 (2000).
5. W. S. B. Paterson, N. Reeh, *Nature* 414, 60 (2001).
6. G. M. Flato *et al.*, *Clim. Dyn.* 23, 229 (2004).
7. J. Brigham-Grette, D. M. Hopkins, *Quaternary Res.* 43, 154 (1995).
8. A. V. Lozhkin, P. M. Anderson, *Quaternary Res.* 43, 147 (1995).
9. M. E. Edwards, T. D. Hamilton, S. A. Elias, N. H. Bigelow, A. P. Krumhardt, *Arctic Antarctic Alpine Res.* 35, 460 (2003).
10. D. Raynaud, J. A. Chappellaz, C. Ritz, P. Matinerie, *J. Geophys. Res.* 102, 26607 (1997).
11. R. M. Koerner, *Science* 244, 964 (1989).
12. O. Bennike, J. Böcher, *Boreas* 23, 479 (1994).
13. C. Hillaire-Marcel, A. De Vernal, G. Bilodeau, A. J. Weaver, *Nature* 410, 1073 (2001).
14. Materials and methods are available as supporting material on *Science* Online.
15. J. T. Kiehl, P. R. Gent, *J. Clim.* 17, 3666 (2004).
16. S. J. Marshall, K. M. Cuffey, *Earth Planet. Sci. Lett.* 179, 73 (2000).
17. J. T. Overpeck *et al.*, *Science* 311, 1747 (2006).
18. M. T. McCulloch, T. Esat, *Chem. Geol.* 169, 107 (2000).
19. M. Montoya, H. von Storch, T. J. Crowley, *J. Clim.* 13, 1057 (2000).
20. J. E. Kutzbach, R. G. Gallimore, P. J. Guetter, *Quaternary Int.* 10–12, 223 (1991).
21. T. J. Crowley, K.-Y. Kim, *Science* 265, 1566 (1994).
22. W. Krabill *et al.*, *Geophys. Res. Lett.* 31, L24402 (2004).
23. R. M. Koerner, D. A. Fisher, *Ann. Glaciology* 35, 19 (2002).
24. K. M. Cuffey, S. J. Marshall, *Nature* 404, 591 (2000).
25. L. Tarasov, W. R. Peltier, *J. Geophys. Res.* 108, 2143 (2003).
26. J. A. Foley, J. E. Kutzbach, M. T. Coe, S. Levis, *Nature* 371, 52 (1994).
27. M. Kerwin *et al.*, *Paleoceanography* 14, 200 (1999).
28. S. P. Harrison, J. E. Kutzbach, I. C. Prentice, P. J. Behling, M. T. Sykes, *Quaternary Res.* 43, 174 (1995).
29. R. J. Stouffer *et al.*, *J. Clim.* 2, 19 (2007).
30. R. B. Alley, P. U. Clark, P. Huybrechts, I. Joughin, *Science* 310, 456 (2005).
31. Circum-Arctic PaleoEnvironments (CAPE) is a program within the International Geosphere-Biosphere Program (IGBP)–Past Global Changes (PAGES); the CAPE-LIG compilation was supported by the NSF–Office of Polar Programs–Arctic System Science (PARCS) and PAGES. The CAPE Last Interglacial Project Members are P. Anderson, O. Bennike, N. Bigelow, J. Brigham-Grette, M. Duvall, M. Edwards, B. Fréchette, S. Funder, S. Johnsen, J. Knies, R. Koerner, A. Lozhkin, G. MacDonald, S. Marshall, J. Matthiessen, G. Miller, M. Montoya, D. Muhs, B. Otto-Bliesner, J. Overpeck, N. Reeh, H. P. Sejrup, C. Turner, and A. Velichko. We thank E. Brady and D. Schimel for helpful discussions, R. Tomas and M. Stevens for figures, and C. Shields for design and running of the simulations. We acknowledge the efforts of a large group of scientists at the NCAR, at several Department of Energy and National Oceanic and Atmospheric Administration labs, and at universities across the United States who contributed to the development of CCSM2. Computing was done at NCAR as part of the Climate Simulation Laboratory. Funding for NCAR and this research was provided by NSF.

Supporting Online Material

www.sciencemag.org/cgi/content/full/311/5768/1751/DC1
Materials and Methods
Figs. S1 to S4
References

A Semi-Empirical Approach to Projecting Future Sea Level Rise

STEFAN RAHMSTORF

A semi-empirical relation is presented that connects global sea level rise to global mean surface temperature. It is proposed that, for time scales relevant to anthropogenic warming, the rate of sea level rise is roughly proportional to the magnitude of warming above the temperatures before the Industrial Age. This holds to good approximation for temperature and sea level changes during the 20th century, with a **proportionality constant** of 3.4 mm/year per °C. When applied to future warming scenarios of the Intergovernmental Panel on Climate Change (IPCC), this relationship results in a projected sea level rise in 2100 of 0.5 to 1.4 m above the 1990 level.

Understanding global sea level changes is a difficult physical problem, because complex mechanisms with different time scales play a role (1), including thermal expansion of water

due to the uptake and penetration of heat into the oceans, input of water into the ocean from glaciers and ice sheets, and changed water storage on land. Ice sheets have the largest potential effect, because their complete melting would result in a global sea level rise of about 70 m. Yet their dynamics are poorly understood, and the key processes that control the response of ice flow to a warming climate are not included in current ice sheet models [for example, melt-

This article first appeared in *Science* (19 January 2007: Vol. 315, no. 368). It has been revised for this edition.

KEY TERM

Two quantities are called proportional if they vary in such a way that one of the quantities is a constant multiple of the other or, equivalently, if they have a constant ratio. This ratio is called the proportionality constant, and it can be determined from empirical data.

Semi-empirical models are based partly on theory and partly on empirical measurements. When they model systems whose processes are not fully understood in theory, modelers may use empirical data to determine the relationship between variables, thus enabling the model to project outcomes without a complete theoretical representation of all elements of the system.

water lubrication of the ice sheet bed (2) or increased ice-stream flow after the removal of buttressing ice shelves (3)]. Large uncertainties exist even in the projection of thermal expansion, and estimates of the total volume of ice in mountain glaciers and ice caps that are remote from the continental ice sheets are uncertain by a factor of two (4). Finally, there are as yet no published physically based projections of ice loss from glaciers and ice caps fringing Greenland and Antarctica.

For this reason, our capability for calculating future sea level changes in response to a given surface-warming scenario with present physics-based models is very limited, and models are not able to fully reproduce the sea level rise of recent decades. Rates of sea level rise calculated with climate and ice sheet models are generally lower than observed rates. Since 1990, observed sea level has followed the uppermost uncertainty limit of the IPCC Third Assessment Report (AR3), which was constructed by assuming the highest emission scenario combined with the highest climate sensitivity and adding an ad hoc amount of sea level rise for "ice sheet uncertainty" (1).

While process-based physical models of sea level rise are not yet mature, **semi-empirical**

models can provide a pragmatic alternative to estimate the sea level response. This is also the approach taken for predicting tides along coasts (e.g., the well-known tide tables), where the driver (tidal forces) is known, but the calculation of the sea level response from first principles is so complex that semi-empirical relationships perform better. Likewise, with current and future sea level rise, the driver is known [global warming (1)], but the computation of the link between the driver and the response from first principles remains elusive. Here, we will explore a semi-empirical method for estimating sea level rise.

We will use the global average near-surface air temperature as a driver; it is the standard diagnostic used to describe global warming. Figure 1 shows a schematic response to a step-function increase in temperature, after climate and sea level parameters were at equilibrium. We expect sea level to rise as the ocean takes up heat and ice starts to melt, until (asymptotically) a new **equilibrium sea level** is reached. Paleoclimatic data suggest that changes in the final equilibrium level may be very large: sea level at the **Last Glacial Maximum**, about 20,000 years ago, was 120 m lower than the current level, whereas global mean temperature was 4°C to 7°C lower (5, 6). Three million years ago, during the Pliocene, the average climate

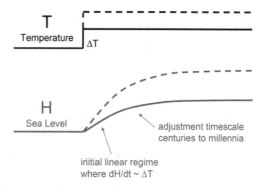

FIGURE 1. Schematic of the response of sea level to a temperature change. The solid line and the dashed line indicate two examples with different amplitude of temperature change.

was about 2°C to 3°C warmer, and sea level was 25 to 35 m higher [7] than today's values. These data suggest changes in sea level on the order of 10 to 30 m per °C.

The initial rate of rise is expected to be proportional to the temperature increase

$$dH/dt = a(T - T_0) \qquad (1)$$

where H is the global mean sea level, t is time, a is the proportionality constant, T is the global mean temperature, and T_0 is the previous equilibrium temperature value. The equilibration time scale is expected to be on the order of millennia. Even if the exact shape of the time evolution $H(t)$ is not known, we can approximate it by assuming a linear increase in the early phase; the long time scales of the relevant processes give us hope that this linear approximation may be valid for a few centuries. As long as this approximation holds, the sea level rise above the previous equilibrium state can be computed as

$$H(t) = a \int_{t0}^{t} (T(t') - T_0) \, dt' \qquad (2)$$

where t' is the time variable.

We test this relationship with observed data sets of global sea level [8] and temperature [combined land and ocean temperatures obtained from NASA (9)] for the period 1880–2001, which is the time of overlap for both series. A highly significant correlation of global temperature and the rate of sea level rise is found ($r = 0.88$, $P = 1.6 \times 10^{-8}$) (Fig. 2) with a slope of $a = 3.4$ mm/year per °C. If we divide the magnitude of equilibrium sea level changes that are suggested by paleoclimatic data [5–7] by this rate of rise, we obtain a time scale of 3000 to 9000 years, which supports the long equilibration time scale of sea level changes.

The baseline temperature T_0, at which sea level rise is zero, is 0.5°C below the mean temperature of the period 1951–1980. This result is consistent with proxy estimates of temperatures in the centuries preceding the modern warming [10], confirming that temperature and sea level were not far from equilibrium before this modern warming began. This is consistent with the time scale estimated above and the relatively stable climate of the Holocene (the past 10,000 years).

In Figure 3, we compare the time evolution of global mean temperature, converted to a "hindcast" rate of sea level rise according to Equation 1, with the observed rate of sea level rise. This comparison shows a close correspondence of the two rates over the 20th century. Like global

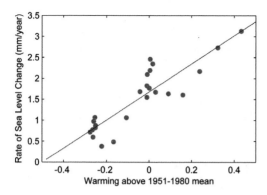

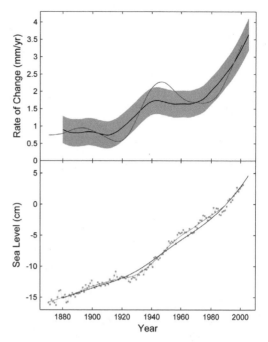

FIGURE 2. Correlation of temperature and the rate of sea level rise for the period 1881–2001. The dashed line indicates the linear fit. Both temperature and sea level curves were smoothed by computing nonlinear trend lines, with an embedding period of 15 years (14). The rate of sea level change is the time derivative of this smoothed sea-level curve, which is shown in Fig. 3. Data were binned in five-year averages to illustrate this correlation.

FIGURE 3. (Top) Rate of sea level rise obtained from tide gauge observations (light blue line, smoothed as described in the Fig. 2 legend) and computed from global mean temperature from Eq. 1 (dark blue line). The light blue band indicates the statistical error (one SD) of the simple linear prediction (15). (Bottom) Sea level relative to 1990 obtained from observations (red line, smoothed as described in the Fig. 2 legend) and computed from global mean temperature from Eq. 2 (blue line). The red squares mark the unsmoothed, annual sea level data.

temperature evolution, the rate of sea level rise increases in two major phases: before 1940 and again after about 1980. It is this figure, that most clearly demonstrates the validity of Equation 1. Accordingly, the sea level that was computed by integrating global temperature with the use of Equation 2 is in excellent agreement with the observed sea level (Fig. 3), with differences always well below 1 cm.

We can explore the consequences of this semi-empirical relationship for future sea levels (Fig. 4), using the range of 21st-century temperature scenarios of the IPCC (1) as input into Equation 2. These scenarios, which span a range of temperature increase from 1.4°C to 5.8°C between 1990 and 2100, lead to a best estimate of sea level rise of 55 to 125 cm over this period. By including the statistical error of the fit shown in Figure 2 (one SD), the range is extended from 50 to 140 cm. These numbers are significantly higher than the model-based estimates of the IPCC for the same set of temperature scenarios, which gave a range from 21 to 70 cm (or from 9 to 88 cm, if the ad hoc term for ice sheet

uncertainty is included). These semi-empirical scenarios smoothly join with the observed trend in 1990 and are in good agreement with it during the period of overlap.

We checked that this analysis is robust within a wide range of embedding periods (i.e., smoothing) of the observational time series. The slope found in Figure 2 varies between 3.2 and 3.5 mm/year per °C for any embedding period between 2 and 17 years, causing only minor variations in the projected sea level. For short embedding periods (about five years), the rate of sea level rise (Fig. 3, top) closely resembles

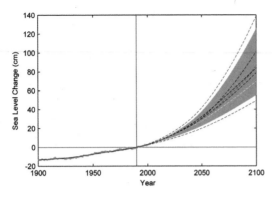

FIGURE 4. Past sea level and sea level projections from 1990 to 2100 based on global mean temperature projections of the IPCC AR3. The gray uncertainty range spans the range of temperature rise of 1.4°C to 5.8°C, having been combined with the best statistical fit shown in Fig. 2. The dashed gray lines show the added uncertainty due to the statistical error of the fit of Fig. 2. Colored dashed lines are the individual scenarios as shown in (1); the light blue line is the A1FI scenario, and the yellow line is the B1 scenario.

that shown in (8) with large short-term fluctuations. For embedding dimensions longer than 17 years, the slope starts to decline because the acceleration of sea level rise since 1980 (Fig. 3) is then progressively lost by excessive smoothing. For very long embedding periods (30 years), the rate of sea level rise becomes rather flat, such as that shown in (11).

The linear approximation (Eq. 1) is only a simplistic first-order approximation to a number of complex processes with different time scales. The statistical error included in Figure 4 does not include any systematic error that arises if the linear relationship breaks down during the forecast period. We can test for this systematic error using climate models, if only for the thermal expansion component of sea level rise that these models capture. For this test, we used the CLIMBER-3α climate model (12), which uses a simplified atmosphere model coupled to a three-dimensional general circulation ocean model with free surface (i.e., that vertically adjusts).

KEY TERM

Smoothing **is a process by which data points are averaged with their neighbors in a series, for the purpose of capturing longer-term trends while filtering out short-term fluctuations.**

We used a model experiment initialized from an equilibrium state of the coupled system in the year 1750 and, with historical radiative forcing, forced changes until the year 2000. After 2000, the model was forced with the IPCC A1FI scenario. The global mean temperature increases by 0.8°C in the 20th century and by 5.0°C from 1990 to 2100 in this experiment.

Temperature and sea level rise data from this model for the time period 1880–2000 were treated as the observational data in the analysis presented above, and graphs corresponding to Figures 2 and 3 look similar to those derived from the observational data (see figs. S1 and S2 online). The slope found is only 1.6 mm/year per °C (i.e., half of the observed slope) because only the thermal expansion component is modeled. Using the semi-empirical relation as fitted to the period 1880–2000, we predicted the sea level for the 21st century (see fig. S3 online). Up to the year 2075, this predicted sea level remains within 5 cm of the actual (modeled) sea level. By the year 2100, the predicted level is 51 cm, whereas the actual (modeled) level is 39 cm above that of 1990 (i.e., the semi-empirical formula overpredicts sea level by 12 cm).

For the continental ice component of sea level rise, we do not have good models to test how the linear approximation performs, although the approximation is frequently used by glaciologists ("degree-days scheme"). Given the dynamical response of ice sheets observed in recent decades and their growing contribution to overall sea level rise, this approximation may not be robust. The ice sheets may respond more strongly to temperature in the 21st century than would be suggested by a linear fit to the 20th century data, if time-lagged positive feedbacks come into play (e.g., bed lubrication, loss of buttressing ice shelves, and ocean warming at the grounding line of ice streams). On the other hand, many small mountain glaciers may disappear within this century and cease to contribute to sea level rise. It is therefore difficult to say whether the linear assumption overall leads to an over- or underestimation of future sea level. Occam's razor suggests that it is prudent to accept the linear assumption as reasonable, although it should be kept in mind that a large uncertainty exists, which is not fully captured in the range shown in Figure 4.

Regarding the lowest plausible limit to sea level rise, a possible assumption may be that the rate shown in Figure 3 stops increasing within a few years (although it is difficult to see a physical reason for this) and settles at a constant value of 3.5 mm/year. This implies a sea level rise of 38 cm from 1990 to 2100. Any lower value would require that the rate of sea level rise drop despite rising temperature, reversing the relationship found in Figure 2.

Although a full physical understanding of sea level rise is lacking, the uncertainty in future sea level rise is probably larger than previously estimated. A rise of over 1 m by 2100 for strong warming scenarios cannot be ruled out, because all that such a rise would require is that the linear relation of the rate of sea level rise and temperature, which was found to be valid in the 20th century, remain valid in the 21st century. On the other hand, very low sea level rise values as reported in the IPCC AR3 now appear rather implausible in the light of the observational data.

The possibility of a faster sea level rise needs to be considered when planning adaptation measures, such as coastal defenses, or mitigation measures designed to keep future sea level rise within certain limits [for example, the 1 m long-term limit proposed by the German Advisory Council on Global Change (13, 16)].

References and Notes

1. J. T. Houghton et al., Eds., Climate Change 2001: The Scientific Basis (Cambridge Univ. Press, Cambridge, 2001).

2. H. J. Zwally et al., Science 297, 218 (2002).

3. E. Rignot et al., Geophys. Res. Lett. 31, 18 (2004).

4. S. C. B. Raper, R. J. Braithwaite, Nature 439, 311 (2006).

5. C. Waelbroeck et al., Quat. Sci. Rev. 21, 295 (2002).

6. T. Schneider von Deimling, A. Ganopolski, H. Held, S. Rahmstorf, Geophys. Res. Lett. 33, L14709 (2006).

7. H. J. Dowsett et al., Global Planet. Change 9, 169 (1994).

8. J. A. Church, N. J. White, Geophys. Res. Lett. 33, L01602 (2006); http://www.sciencemag.org/cgi/external_ref?access_num=10.1029/2005GL024826&link_type=DOI.

9. J. Hansen et al., J. Geophys. Res. Atmos. 106, 23947 (2001).

10. P. D. Jones, M. E. Mann, Rev. Geophys. 42, RG2002 (2004).

11. S. Jevrejeva, A. Grinsted, J. C. Moore, S. Holgate, J. Geophys. Res. 111, 09012 (2006).

12. M. Montoya et al., Clim. Dyn. 25, 237 (2005).

13. German Advisory Council on Global Change, The Future Oceans: Warming Up, Rising High, Turning Sour (Wissenschaftlicher Beirat der Bundesregierung Globale Umweltveränderungen Special Report, Berlin, 2006); www.wbgu.de/wbgu_sn2006_en.pdf.

14. J. C. Moore, A. Grinsted, S. Jevrejeva, Eos 86, 226 (2005).

15. The statistical error was calculated by means of the Matlab function "polyval" for the linear fit shown in Fig. 2.

16. The author thanks J. Church for providing the observational data and M. Stöckmann for the model data. They as well as J. Gregory and B. Hare are thanked for valuable discussions.

Supporting Online Material

www.sciencemag.org/cgi/content/full/1135456/DC1
Figs. S1 to S3

Prioritizing Climate Change Adaptation Needs for Food Security in 2030

DAVID B. LOBELL, MARSHALL B. BURKE, CLAUDIA
TEBALDI, MICHAEL D. MASTRANDREA, WALTER P. FALCON,
ROSAMOND L. NAYLOR

Investments aimed at improving agricultural adaptation to climate change inevitably favor some crops and regions over others. An analysis of climate risks for crops in 12 food-insecure regions was conducted to identify adaptation priorities, based on statistical crop models and climate projections for 2030 from 20 general circulation models (GCMs). Results indicate South Asia and Southern Africa as two regions that, without sufficient adaptation measures, will likely suffer negative impacts on several crops that are important to large food-insecure human populations. We also find that uncertainties vary widely by crop, and therefore priorities will depend on the risk attitudes of investment institutions.

This article first appeared in *Science* (1 February 2008: Vol. 319, no. 5863). It has been revised for this edition.

Adaptation is a key factor that will shape the future severity of climate change impacts on food production (1). Although relatively inexpensive changes, such as shifting planting dates or switching to an existing crop variety, may moderate negative impacts, the biggest benefits will likely result from more costly measures, including the development of new crop varieties and expansion of irrigation (2). These adaptations will require substantial investments by farmers, governments, scientists, and development organizations, all of whom face many other demands on their resources. Prioritization of investment needs, such as through the identification of "climate risk hot spots" (3), is therefore a critical issue but has received limited attention to date.

We consider three components to be essential to any prioritization approach: (i) selection of a time scale over which impacts are most relevant to investment decisions, (ii) a clear definition of criteria used for prioritization, and (iii) an ability to evaluate these criteria across a suite of crops and regions. Here, we focus on food security impacts by 2030—a time period most relevant to large agricultural investments, which typically take 15 to 30 years to realize full returns (4, 5).

We consider several different criteria for this time scale. First is the importance of the crop to a region's food-insecure human population [hunger importance (HI)]. Second is the median projected impact of climate change on a crop's production by 2030 (indicated by C50), assuming no adaptation. For this analysis, we generate multiple (i.e., 100) projections of impacts based on different models of climate change and crop response, in order to capture relevant uncertainties. The projections are then ranked, and the average of the 50th and 51st values are used as the median. A third criterion is the fifth percentile of projected impacts by 2030 (where C05 indicates the fifth value of the ranked projections), which we use to represent

the lower tail, or "worst case," among the projections. Finally, we consider the 95th percentile of projected impacts by 2030 (where C95 indicates the 95th value of the ranked projections), which we use to represent the upper tail, or "best case," among the projections.

We first identified 12 major food-insecure regions, each of which (i) comprise groups of countries with broadly similar diets and agricultural production systems and (ii) contain a notable share of the world's malnourished individuals as estimated by the Food and Agriculture Organization (FAO) (Table 1; see fig. S1 online for details on regions). For each region, we computed the HI value for each crop by multiplying the number of malnourished individuals by the crop's percent contribution to average per-capita calorie consumption [see supporting online material (SOM) Text S1 and table S1]. A hunger importance ranking (HIR) was then generated by ranking the HI values for all crop-by-region combinations. Rice, maize, and wheat contribute roughly half of the calories currently consumed by the world's poor and

TABLE 1. Regions evaluated in this study and selected summary statistics. Countries within each region are indicated in the SOM.

Region	Code	Malnourished			
		Millions of people	Crops modeled	Crops with significant model[*]	World total (%)
South Asia	SAS	262.6	30.1%	9	7
China	CHI	158.5	18.2%	7	2
Southeast Asia	SEA	109.7	12.6%	7	4
East Africa	EAF	79.0	9.1%	10	2
Central Africa	CAF	47.6	5.5%	8	0
Southern Africa	SAF	33.3	3.8%	8	6
West Africa	WAF	27.5	3.2%	8	2
Central America and Caribbean	CAC	25.4	2.9%	5	2
Sahel	SAH	24.9	2.9%	7	7
West Asia	WAS	21.9	2.5%	10	4
Andean region	AND	21.4	2.5%	9	3
Brazil	BRA	13.5	1.6%	6	4
Total	ALL	825.3	94.7%	94	43

[*] A model was judged significant if it explained more than 14% of variance in yield or production ($R^2 > 0.14$). This threshold was based on the 95th percentile of the R^2 statistic from a Monte Carlo experiment, which computed 1000 multiple regression models for a randomly generated 42-year time series with two random predictor variables.

only 31% of the calories consumed by those in sub-Saharan Africa, illustrating the importance of considering additional crops in food security assessments. The use of projected malnourished populations in 2030 rather than current population values had a very small influence on the rankings (see table S2 online).

Several options exist for evaluating climate change impacts across a suite of crops and regions (see SOM Text S2 online). We used data sets on historical crop harvests (6), monthly temperatures and precipitation, and maps of crop locations to develop statistical crop models for 94 crop-by-region combinations spanning the 12 study regions (see SOM Text S3 online;

results summarized in Table 1). Of these combinations, 46% (43) exhibited a statistically significant model ($P < 0.05$), and 22% (21) had a model R^2 of at least 0.3. As seen in the examples for wheat in South and West Asia (see fig. S3 online), in some cases, the model's strength came primarily from a (typically negative) temperature effect on yield, whereas, in other cases, a (typically positive) rainfall effect provided most of the explanatory power.

The crop temperature sensitivities estimated by the statistical models were compared with corresponding values from previous studies that relied on established process-based models within the same regions (see SOM Text S4 online). Our statistical estimates generally overlapped the lower end of the range of previous estimates, indicating that impacts estimated by the statistical models may be considered conservative but in reasonable agreement with estimates from process-based approaches.

To project climate changes for the crop regions, along with their uncertainties, we used output from 20 GCMs that have contributed to the World Climate Research Programme's

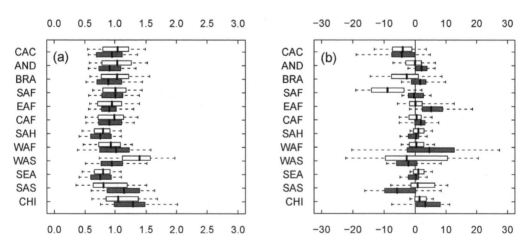

FIGURE 1. Summary of projected (a) temperature (°C) and (b) precipitation (%) changes for 2030 (the averages from 2020–2039 relative to those from 1980–1999) based on output from 20 GCMs and three emission scenarios. Gray boxes show DJF averages, and white boxes show JJA averages. Dashed lines extend from 5th to 95th percentile of projections, boxes extend from 25th to 75th percentile, and the middle vertical line within each box indicates the median projection.

*Science*NOW Daily News, 23 July 2007

THE POWER TO INFLUENCE SHOWERS
Phil Berardelli

A new study has shown for the first time that humans have been altering rainfall patterns across various parts of the globe for the past century. The research could help scientists predict future rainfall patterns within geographic regions and allow nations to prepare better for changing weather.

For more than two decades, the world's climate scientists have been building a case that the by-products of human activity—particularly greenhouse gases—are warming the planet. At the same time, scientists have been investigating whether those greenhouse gases are affecting global precipitation, perhaps causing the severe droughts that have afflicted the U.S. Southwest and Africa's Sahel region for years (*Science*NOW, 10 October 2003). So far, however, computer models have only *suggested* that rainfall patterns have changed due to the influence of *Homo sapiens*.

Now, an international group of scientists has provided the first empirical evidence that human activity really is having an effect on precipitation. The researchers cross-checked detailed rainfall observations going back over 80 years against a new and extensive array of climate change simulations that take into account two types of emissions related to human activity: greenhouse gases and sulfate aerosols. They then broke down those data into broad latitudinal bands circling the planet and restricted the analyses to rainfall over land. After comparing observational data with the results of 92 separate simulations, the team has concluded that humans are indeed altering rain patterns in three latitudinal regions. Specifically, the computer models show that human activity has added up to two-thirds of the extra rain observed in the northern temperate regions, including Canada, the United States, Europe, and Russia; removed up to a third of the rain that has disappeared from the northern tropics and subtropics, including Mexico and Africa's Sahara and Sahel regions; and added nearly all of the additional precipitation in the southern tropics and subtropics, including Brazil, Southern Africa, and Indonesia.

The amounts are not trivial. In the case of the southern tropics and subtropics, up to 82 mm more rain has fallen per year over the past century, the team reports in *Nature*. So, in addition to changing temperatures, changing precipitation will present environmental and economic challenges to humans living in the affected areas, the researchers say. But other researchers wonder whether the team is placing too much blame on humans. One potential problem is that no one knows what long-term natural patterns of drought and recovery look like, says climate scientist David Battisti of the University of Washington, Seattle, so linking the patterns to human activity may be premature. Another difficulty could arise from including extremely arid regions like . the Sahel in the latitudinal bands, says researcher Michael Glantz of the National Center for Atmospheric Research in Boulder, Colorado, because rainfall in such areas can be so erratic that "it's difficult to separate the signal from the noise."

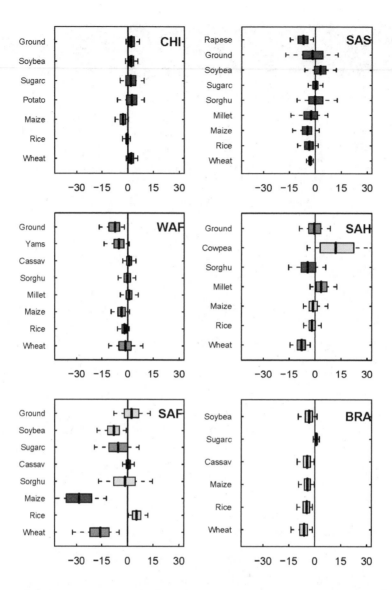

FIGURE 2. Probabilistic projections of production impacts in 2030 from climate change (expressed as a percentage of 1998–2002 average yields). Red, orange, and yellow indicate a HIR of 1 to 30 (more important), 31 to 60 (important), and 61 to 94 (less important), respectively. Dashed lines extend from 5th to 95th percentile of projections, boxes extend from 25th to 75th percentile, and the middle vertical line within each box indicates the median projection. Region codes are defined in Table 1.

Coupled Model Intercomparison Project phase 3 (WCRP CMIP3) (7). Median projections of average temperature change from 1980–2000 to 2020–2040 were roughly 1.0°C in most regions, with few models projecting less than 0.5°C warming in any season and some models warming by as much as 2.0°C (Fig. 1a). In contrast to the unanimous warming, models were mixed in the direction of simulated precipitation change. All regions had at least one model with positive and one model with negative projected precipitation changes, with median projections ranging from about –10% to +5% (Fig. 1b). Some relevant tendencies of current GCMs, as noted

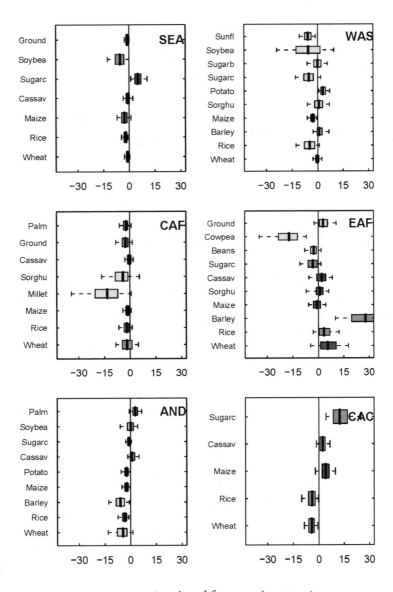

FIGURE 2. *(continued from previous page)*

in (*8*), are toward precipitation decreases during December to February (DJF) in South Asia and Central America, precipitation decreases in June to August (JJA) in Southern Africa, Central America, and Brazil, and precipitation increases in DJF in East Africa.

We estimated a probability distribution of production changes for 2030 (the average from 2020–2039 relative to that from 1980–1999) for each crop, using a Monte Carlo procedure that propagated both climate and crop uncer-

tainties (*9*). To facilitate comparison between crops and regions, we expressed production changes for all crops as a percentage of average values for 1998–2002. The impact projections are summarized in Figure 2.

For simplicity, we consider three general classes of projections. First, several projections (e.g., Southern Africa maize and wheat) are consistently negative, with an estimated 95% or greater chance that climate changes will harm crop production in the absence of adaptation

(C95 < 0). These cases generally arise from a strong dependence of historical production variations on temperature, combined with projected warming large enough to overwhelm the uncertain impacts of precipitation changes.

Second, there are many cases with large uncertainties, with model impacts ranging from substantially negative to positive (e.g., South Asia groundnut, Southern Africa sorghum). These cases usually arise from a relatively strong dependence of historical production on rainfall, combined with large uncertainties in future precipitation changes. More precise projections of precipitation would therefore be particularly useful to reduce impact uncertainties in these situations. Large uncertainties also arise in some cases (e.g., cowpea in East Africa) from an estimated production response to historical temperature that is strongly negative but also highly uncertain.

Finally, there are many cases characterized by a narrow 90% confidence interval of impacts within ±5% of zero. In a few cases, such as wheat in West Asia, this reflects a strong effect of historical rainfall variations (see fig. S1 online), combined with a relatively narrow range of rainfall projections during the growing season (Fig. 1; West Asia wheat is grown in DJF). In most cases, such as cassava in West Africa, the narrow confidence intervals result from a relatively weak relationship between historical production and growing-season climate. Therefore, we can only say that the likely impacts appear small,

given the current data sets and models used to describe crop responses to climate. In cases with low model R^2, approaches other than the FAO-based regression models used here may be more appropriate.

Based on the above projections, we identified a small subset of crops that met different prioritization criteria (Table 2). First, crops were separated into groups of "more important" (HIR = 1 to 30), "important" (HIR = 31 to 60), and "less important" (HIR = 61 to 94). Within each category, we identified crops below three thresholds: the first corresponding to instances where at least 5% of the models predicted greater than 10% loss of production (C05 < −10%), the second to where at least half the models projected greater than a 5% production loss (C50 < −5%), and the third to where at least 95% of the models predicted some production loss (C95 < 0%).

Although several crops met more than one of these criteria, such as maize in Southern Africa and rapeseed in South Asia, the varying estimates of uncertainty for different crops, in general, resulted in noticeable differences when prioritizing crops on the basis of the three different thresholds (Table 2). For example, a relatively weak relationship was found between values at the two tails of the projection distributions—C05 and C95—across all crops (see fig. S4 online). This result indicates a need to explicitly consider uncertainty and risk attitudes when setting priorities, which is an issue that has received limited attention (10).

Because attitudes toward risk differ, and given that impact projections for some crops are more uncertain than those for other crops, various institutions might derive different priorities from the results in Table 2. For example, one set of institutions might wish to focus on those cases where negative impacts are most likely to occur, in order to maximize the likelihood that investments will generate some benefits. By this criterion (C95 < 0%), South Asia wheat, Southeast Asia rice, and Southern Africa maize appear as the most important crops in need of adaptation investments.

Projecting the Future

TABLE 2. Crop priority lists based on different criteria. C05 = 5th percentile of projected impacts (5th lowest out of 100 projections); C50 = 50th percentile (median); C95 = 95th percentile. Results are shown only for the HIR = 1 to 30 and HIR = 31 to 60 categories.

HIR value	Criterion	Crops
1 to 30	C05 < -10%	South Asia millet, groundnut, rapeseed; Sahel sorghum; Southern Africa maize
	C50 < -5%	South Asia rapeseed; Southern Africa maize
	C95 < 0%	South Asia wheat; Southeast Asia rice; Southern Africa maize
31 to 60	C05 < -10%	Southeast Asia soybean; West Asia rice; Western Africa wheat, yams, groundnut; Sahel wheat; East Africa sugarcane; Southern Africa wheat, sugarcane; Brazil wheat, rice; Andean Region wheat; Central America rice
	C50 < -5%	Southeast Asia soybean; West Asia rice; Western Africa yams, groundnut; Sahel wheat; Southern Africa wheat, sugarcane; Brazil wheat
	C95 < 0%	Western Africa groundnut; Sahel wheat; Southern Africa wheat; Brazil wheat, rice; Central America wheat, rice

Others might argue that adaptation activities that do not account for worst-case projections will be inadequate in the face of low-probability, high-consequence climate impacts: that is to say, investments should target those crops and regions for which some models predict very negative outcomes. A different subset of crops is identified for this criterion (C05 < −10%), with several South Asian crops, Sahel sorghum, and (again) Southern Africa maize appearing as the most in need of attention.

Either of these risk attitudes could be applied with an explicit regional focus. For a sub-Saharan African institution interested in investing where negative impacts are most likely to occur [where median impact projections are substantially negative (C50 < −5%) or where most climate models agree that negative impacts are likely to occur (C95 < 0%)], priority investments would include Southern Africa maize, wheat, and sugarcane, Western Africa yams and groundnut, and Sahel wheat.

Despite the many assumptions and uncertainties associated with the crop and climate models used (see SOM Text S5 online), the above analysis points to many cases where food security is clearly threatened by climate change in the relatively near term. The importance of adaptation in South Asia and Southern Africa appears particularly robust, because crops in these regions appear for all criteria considered

here (Table 2). The results also highlight several regions (e.g., Central Africa) where climate-yield relationships are poorly captured by current data sets, and therefore future work in this regard is needed to inform adaptation efforts.

Impacts will likely vary substantially within individual regions according to differences in biophysical resources, management, and other factors. The broad-scale analysis presented here was intended only to identify major areas of concern, and further studies at finer spatial scales are needed to resolve local hot spots within regions. Consideration of other social and technological aspects of vulnerability, such as the existing adaptive capacity in a region or the difficulty of making adaptations for specific cropping systems, should also be integrated into prioritization efforts. Although we do not attempt to identify the particular adaptation strategies that should be pursued, we note that, in some regions, switching from highly impacted to less-impacted crops may be one viable adaptation option. In this case, the identification of less-impacted crops is another valuable outcome of a comprehensive approach that simultaneously considers all crops relevant to the food-insecure (11).

References and Notes

1. W. Easterling et al., in Climate Change 2007: Impacts, Adaptation and Vulnerability. Contribution of Working Group II to the Fourth Assessment Report of

the *Intergovernmental Panel on Climate Change* (Cambridge Univ. Press, Cambridge, 2007), pp. 273–313.

2. C. Rosenzweig, M. L. Parry, *Nature* 367, 133 (1994).

3. I. Burton, M. van Aalst, *Look Before You Leap: A Risk Management Approach for Incorporating Climate Change Adaptation in World Bank Operations* (World Bank, Washington, DC, 2004).

4. J. M. Alston, C. Chan-Kang, M. C. Marra, P. G. Pardey, T. J. Wyatt, *A Meta-Analysis of Rates of Return to Agricultural R&D: Ex Pede Herculem?* (International Food Policy Research Institute, Washington, DC, 2000).

5. J. Reilly, D. Schimmelpfennig, *Clim. Change* 45, 253 (2000).

6. We used FAO data on national crop production and area, which include quantities consumed or used by the producers in addition to those sold on the market.

7. Model simulations under three SRES (Special Report on Emissions Scenarios) emission scenarios corresponding to relatively low (B1), medium (A1b), and high (A2) emission trajectories were used. Although the mean projections for the emission scenarios exhibit very small differences out to 2030, the use of three scenarios provided a larger sample of simulations with which to assess climate uncertainty. For all simulations, average monthly output for 1980–1999 was subtracted from that of 2020–2039 to compute monthly changes in temperature and precipitation.

8. J. H. Christensen *et al.*, in *Climate Change 2007: The Physical Science Basis. Contribution of Working Group I to the Fourth Assessment Report of the Intergovernmental Panel on Climate Change*, S. Solomon *et al.*, Eds. (Cambridge Univ. Press, Cambridge, 2007), pp. 847–940.

9. Namely, the crop regression model was fit with a bootstrap sample from the historical data, and the coefficients from the regression model were then multiplied by projected changes in average temperature and precipitation, which were randomly selected from the CMIP3 database. This process was repeated 100 times for each crop. Bootstrap resampling is a common approach to estimate uncertainty in regression coefficients, although this addresses only the component of model uncertainty that arises from a finite historical sample and not the potential uncertainty from structural errors in the model. Similarly, the representation of climate uncertainty by equally weighting all available GCMs is a common approach but could potentially be improved, such as by weighting models according to their agreement with the model consensus and with historical observations. Nonetheless, the resulting probability distributions incorporate reasonable measures of both climate and crop uncertainty, and thus they should fairly reflect both the absolute and relative level of uncertainties between crops.

10. B. Smit *et al.*, in *Climate Change 2001: Impacts, Adaptation and Vulnerability. Contribution of Working Group II to the Third Assessment Report of the Intergovernmental Panel on Climate Change* (Cambridge Univ. Press, Cambridge, 2001), pp. 877–912.

11. We thank D. Battisti, C. Field, and three anonymous reviewers for helpful comments. D.B.L. was supported by a Lawrence Fellowship from LLNL. Part of this work was performed under the auspices of the U.S. Department of Energy (DOE) by LLNL under contract DE-AC52-07NA27344. We acknowledge the modeling groups, the Program for Climate Model Diagnosis and Intercomparison, and the WCRP's Working Group on Coupled Modelling for their roles in making available the WCRP CMIP3 multimodel data set. Support of this data set is provided by the Office of Science, DOE.

Supporting Online Material

www.sciencemag.org/cgi/content/full/319/5863/607/DC1
SOM Text S1 to S5
Figs. S1 to S5
Tables S1 to S3
References

Critical Assumptions in the Stern Review on Climate Change

WILLIAM NORDHAUS

I
n November 2006, the British government presented a comprehensive study on the economics of climate change (1), the Stern Review. It painted a dark picture for the globe: "[I]f we don't act, the overall costs and risks of climate change will be equivalent to losing at least 5% of global GDP [gross domestic product] each year, now and forever. If a wider range of risks and impacts is taken into account, the estimates of damage could rise to 20% of GDP or more." The Stern Review recommended urgent, immediate, and sharp reductions in greenhouse-gas emissions.

These findings differ markedly from economic models that calculate least-cost emissions paths to stabilize concentrations or paths that balance the costs and benefits of emis-

This article first appeared in *Science* (13 July 2007: Vol. 317, no. 5835). It has been revised for this edition.

sions reductions. Mainstream economic models definitely find it economically beneficial to take steps today to slow warming, but efficient policies generally involve modest rates of emissions reductions in the near term, followed by sharp reductions in the medium and long term (2–5).

A standard way of showing the stringency of policies is to calculate the "carbon tax," or penalty on carbon emissions. A recent study by the author estimates an optimal carbon tax for 2005 of about $30 per ton carbon in today's prices, rising to $85 by the mid-21st century and further increasing after that (5). A similar carbon price has been found in studies that estimate the least-cost path to stabilize carbon dioxide concentrations at two times preindustrial levels (2). The sharply rising carbon tax reflects initially low, but rising, emissions reduction rates. We call this the climate-policy ramp, in which policies to slow global warming increasingly tighten or ramp up over time. A $30 carbon

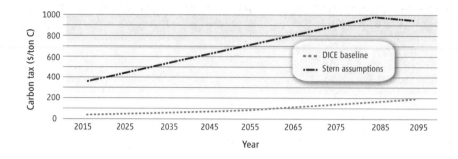

IMAGE 1. Comparing the optimal carbon tax under alternative discounting assumptions. The Dynamic Integrated model of Climate and the Economy (DICE model) (5) integrates the economic costs and benefits of greenhouse gases (GHG) reductions with a simple dynamic representation of the scientific and economic links of output, emissions, concentrations, and climate change. The DICE model is designed to choose levels of investment in tangible capital and in GHG reductions that maximize economic welfare. It calculates the optimal carbon tax as the price of carbon emissions that will balance the incremental costs of abating carbon emissions with the incremental benefits of lower future damages from climate change. Using the DICE model to optimize climate policy leads to an optimal carbon tax in 2005 of around $30 per ton carbon (shown here as "DICE baseline"). If we substitute the Stern Review's assumptions about time discounting and the consumption elasticity into the DICE model, the calculated optimal carbon tax is much higher and rises much more rapidly (shown as "Stern assumptions").

tax may appear to be a modest target, but it is at least 10 times the current globally averaged carbon tax implicit in the Kyoto Protocol (shown as Stern assumptions).

What is the logic of the ramp? In a world where capital is productive and damages are far in the future (see chart above), the highest-return investments today are primarily in tangible, technological, and human capital. In the coming decades, damages are predicted to rise relative to output. As that occurs, it becomes efficient to shift investments toward more intensive emissions reductions and the accompanying higher carbon taxes. The exact timing of emissions reductions depends on details of costs, damages, learning, and the extent to which climate change and damages are nonlinear and irreversible.

The Stern Review proposes to move the timetable for emissions reductions sharply forward. It suggests global emissions reductions of between 30% and 70% over the next two decades, objectives consistent with a carbon tax of about $300

per ton today, or about 10 times the level suggested by standard economic models.

Given that the Stern Review embraces traditional economic techniques such as those described in·(2–5), how does it get such different results and strategies? Having analyzed the Stern Review in (6) (which also contains a list of recent analyses), I find that the difference stems almost entirely from its technique for calculating **discount rates** and only marginally on new science or economics. The reasoning has questionable foundations in terms of its ethical assumptions and also leads to economic results that are inconsistent with market data.

Some background on growth economics and discounting concepts is necessary to understand the debate. In choosing among alternative trajectories for emissions reductions, the key economic variable is the real return on capital, r, which measures the net yield on investments in capital, education, and technology. In principle, this is observable in the marketplace. For example, the real pretax return on U.S. corpoa positive discount rate means that the welfare of future generations is reduced or "discounted" compared with nearer generations.

Analyses are sometimes divided between the "descriptive approach," in which assumed discount rates should conform to actual political and economic decisions and prices, and the "prescriptive approach," where discount rates should conform to an ethical ideal, sometimes taken to be very low or even zero. Philosophers and economists have conducted vigorous debates about how to apply discount rates in areas as diverse as economic growth, climate change, energy, nuclear waste, major infrastructure programs, hurricane levees, and reparations for slavery.

The Stern Review takes the prescriptive approach in the extreme, arguing that it is indefensible to make long-term decisions with a positive time discount rate. The actual time discount rate used in the Stern Review is 0.001 per year, which is vaguely justified by estimates of the probability of the extinction of the human race.

The second parameter that determines return on capital is the consumption elasticity, denoted

rate capital over the last four decades has averaged about 0.07 per year. Estimated real returns on human capital range from 0.06 to > 0.20 per year, depending on the country and time period (7). The return on capital is the "discount rate" that enters into the determination of the efficient balance between the cost of emissions reductions today and the benefit of reduced climate damages in the future. A high return on capital tilts the balance toward emissions reductions in the future, whereas a low return tilts reductions toward the present. The Stern Review's economic analysis recommended immediate emissions reductions because its assumptions led to very low assumed real returns on capital.

Where does the return on capital come from? The Stern Review and other analyses of climate economics base the analysis of real returns on the optimal economic growth theory (8, 9). In this framework, the real return on capital is an economic variable that is determined by two normative parameters. The first parameter is the time discount rate, denoted by ρ, which refers to the discount on future "utility" or welfare (not on future goods, like the return on capital). It measures the relative importance in societal decisions of the welfare of future generations relative to that of the current generation. A zero discount rate means that all generations into the indefinite future are treated the same;

SCIENCE IN THE NEWS

*Science*NOW Daily News, 15 October 2007

THE ECONOMICS NOBEL:
GIVING ADAM SMITH A HELPING HAND
Adrian Cho

Scottish philosopher Adam Smith asserted that when everyone acts out of self-interest, everyone will eventually benefit, as if a benevolent "invisible hand" molds the economy. Economists now know that view is naive: they can prove that in some situations, rational people will act in ways that leave everybody a loser. But such dreary outcomes can sometimes be avoided, thanks to work that today earned three Americans the Nobel Prize in Economics.

Leonid Hurwicz of the University of Minnesota, Twin Cities, Eric Maskin of the Institute for Advanced Study in Princeton, New Jersey, and Roger Myerson of the University of Chicago, Illinois, developed "mechanism design theory." Such study aims to find schemes, or "mechanisms," that ensure that acting in self-interest will indeed lead to benefits for all. Today, the theory's applications range from how best to auction broadcast rights and other public resources to contract negotiations and elections.

Mechanism design theory starts with the recognition that unbridled self-interest doesn't always lead to the greater good. For example, suppose the people of a town would benefit if they built a bridge across the river. Everyone is asked to estimate how much the bridge is worth to him personally and chip in accordingly. Rationally, each person benefits by underestimating his stake in the bridge and letting others bear the cost. So for lack of funds, the bridge never gets built, and the whole community suffers.

In the 1960s, Hurwicz pioneered the study of how to avoid such dead ends by fiddling with the rules of such an economic interaction. For example, in the case of the bridge, the amount each person pays could be based on only what others think the bridge should be worth, thus eliminating each person's incentive to lie about its value.

Maskin and Myerson expanded on Hurwicz's work. For example, in 1977, Maskin developed a criterion for determining just when it's possible to find a set of rules that will guide self-interested participants to the desired end. "This sets some boundaries on what mechanism design theory can do," says Massimo Morelli, an economist at Columbia. Starting in the late 1970s, Myerson showed that whenever a mechanism exists, it is also possible to find one that gives participants an incentive to tell the truth.

Relying heavily on game theory, the laureates' work has been largely abstract and formal. Nevertheless, the theory may play a role in confronting perhaps the most complex and pressing problem facing humanity today, climate change, by helping to set up incentives that encourage consumers and countries to minimize greenhouse gas emissions. But first, politicians must identify the specific end they are working toward, Maskin says. "Mechanism design should definitely be pertinent to the problem," he says, "but first we have to decide exactly what we're trying to accomplish."

as η. This parameter represents the aversion to the economic equality among different generations. A low (high) value of η implies that decisions take little (much) heed about whether the future is richer or poorer than the present. Under standard optimal growth theory, if time discounting is low and society cares little about income inequality, then it will save a great deal for the future, and the real return will be low. This is the case assumed by the Stern Review. Alternatively, if either the time discount rate is high or society is averse to inequality, the current savings rate is low and the real return is high.

This relation is captured by the "Ramsey equation" of optimal growth theory (8, 9), in which the long-run equilibrium real return on capital is determined by $r = \rho + \eta g$, where g is the average growth in consumption per capita, ρ is the time discount rate, and η is the consumption elasticity. Using the Stern Review's assumption of $\rho = 0.001$/year and $\eta = 1$, along with its assumed growth rate ($g = 0.013$/year) and a stable population, yields an equilibrium real interest rate of 0.014/year, far below the returns to standard investments. It would also lead to much higher savings rates than today's. This low rate of return is used in the Stern Review without any reference to actual rates of return or savings rates.

The low return also means that future damages are discounted at a low rate, and this helps explain the Stern Review's estimate that the cost of climate change could represent the equivalent of a "20% cut in per-capita consumption, now and forever." When the Stern Review says that there are substantial losses "now," it does not mean "today." In fact, the Stern Review's estimate of the output loss "today" is essentially zero. We can illustrate this using the Stern Review's high-climate scenario with catastrophic and nonmarket impacts. For this case, the mean losses are 0.4% of world output in 2060, 2.9% in 2100, and 13.8% in 2200. This is reported as a loss in "current per-capita consumption" of 14.4%.

How do damages that average about 1% over the next century turn into 14.4% cuts "now and forever"? The answer is that with the low interest rate, the relatively small damages in the next two centuries get overwhelmed by the high damages over the centuries and millennia that follow 2200. In fact, if the Stern Review's methodology is used, more than half of the estimated damages "now and forever" occur after 2800.

What difference would it make if we used assumptions that are consistent with standard returns to capital and savings rates? For example, take the Stern Review's near-zero time discount rate with a high inequality aversion represented by a consumption elasticity of $\eta = 3$. This combination would yield real returns and savings rates close to those observed in today's economy and dramatically different from those shown in the Stern Review. The optimal carbon tax and the social cost of carbon decline by a factor of about 10 relative to these consistent with the Stern Review's assumptions, and the efficient trajectory looks like the policy ramp discussed above. In other words, the Stern Review's alarming findings about damages, as well as its economic rationale, rest on its model parameterization—a low time discount rate and low inequality aversion—that leads to savings rates and real returns that differ greatly from actual market data. If we correct these parameterizations, we get a carbon tax and emissions reductions that look like standard economic models.

The Stern Review's unambiguous conclusions about the need for urgent and immediate action will not survive the substitution of assumptions that are consistent with today's marketplace real interest rates and savings rates. So the central questions about global warming policy—how much, how fast, and how costly—remain open.

References and Notes

1. N. Stern, *The Economics of Climate Change: The Stern Review* (Cambridge Univ. Press, Cambridge, 2007).

2. J. Weyant, Ed., *Energy Econ.* 26, 4 (Special Issue on EMF 19) 501–755 (2004).

3. W. D. Nordhaus, J. Boyer, *Warming the World: Economic Modeling of Global Warming* (MIT Press, Cambridge, MA, 2000).

4. R. S. J. Tol, *Energy Policy* 33, 2064–2074 (2005).

5. W. D. Nordhaus, The Challenge of Global Warming: Economic Models and Environmental Policy (Yale Univ., New Haven, CT, 2007); http://nordhaus.econ .yale.edu/recent_stuff.html.

6. W. D. Nordhaus, *J. Econ. Lit.*, in press; http:// nordhaus.econ.yale.edu/recent_stuff.html.

7. J. Arrow *et al.*, *Climate Change 1995—Economic and Social Dimensions of Climate Change*, J. Bruce, H. Lee, E. Haites, Eds. (Cambridge Univ. Press, Cambridge, 1996), pp. 125–144.

8. F. Ramsey, *Econ. J.* 38, 543 (1928).

9. T. C. Koopmans, *Acad. Sci. Scripta Varia* 28, 1 (1965).

Climate Change: Risk, Ethics, and the Stern Review

NICHOLAS STERN AND CHRIS TAYLOR

A ny thorough analysis of policy on climate change must examine scientific, economic, and political issues and many other relationships and structures and must have ethics at its heart. In a Policy Forum in this issue of *Science*, Nordhaus (1) suggests that our results as described in the Stern Review (2) stem almost entirely from ethical judgments. This is not correct. In addition to revisiting the ethics, we also incorporated the latest science, which tells us that, for a given change in atmospheric concentration, the worst impacts now appear more likely. Further, the science also now gives us a better understanding of probabilities, so we could incorporate explicit risk analysis, largely overlooked in previous studies. It is risk plus ethics that drive our results.

This article first appeared in *Science* (13 July 2007: Vol. 317, no. 5835). It has been revised for this edition.

The most direct way to look at the problem of constructing an economic response to climate change is to look at the individual impacts of climate change alongside the cost of reducing emissions and then to ask whether it is worth paying for mitigation. However, we do not have the kind of information that would enable formally attaching numbers to all consequences, weighting them, and adding them all up with any plausibility. Thus, economists attempt aggregations of impacts and costs using very simplified aggregate modeling and, in the process, throw away much that is of fundamental importance to a balanced judgment.

The central estimate of mitigation costs for stabilizing emissions below 550 parts per million (ppm) carbon dioxide (CO_2) equivalent is 1% of gross domestic product (GDP) per annum (2). The basic question is thus whether it is worth paying 1% of GDP to avoid the additional risks of higher emissions. The modeling in the Stern Review is valuable in identifying some

Sensitivity analysis is an investigation into how projected performance varies with changes in the key assumptions on which the projections are based.

key drivers of costs and benefits in terms of economic modeling approaches, scientific variables, and ethical considerations. However, excessive focus on the narrow aspects of these simplistic models distorts and often exaggerates their role in policy decisions. They cannot substitute for the detailed risk and cost analysis of key effects.

Our sensitivity analysis shows that our main conclusions—that the costs of strong action are less than the costs of the damage avoided by that action—are robust to a range of assumptions.

These assumptions concern (i) model structure and inputs (including population, structure of the damage function, aversion to irreversible consequences, future conditions, and the rise in price of environmental goods relative to consumption goods) and (ii) value judgments (attitudes to risk and inequality, the extent to which future generations matter, and intragenerational income distribution and/or regional equity weighting).

Some credible assumptions about the rate at which climate change will result in damage would lead to cost estimates that are much higher; our modeling approach has been cautious. Some modelers are very optimistic about economic growth and social rates of return for the next centuries. However, they appear to overlook that such rapid growth is likely to lead to greater emissions and, hence, the more rapid onset of climate change.

The ethical approach adopted in our analysis focuses on the ethics of allocation between richer and poorer people and between those born at different times. Ramsey (3) developed the standard social welfare discounting formula $r = \eta g + \rho$, where r is the consumption discount rate, η is the elasticity of the social benefits attained (also called the social marginal utility), g is per-capita consumption growth rate, and ρ is the time discount rate (also called the pure rate of time preference). The equation arises from comparing the social value of a bit of consumption in the future with a unit now and asking how it falls over time, the definition of a discount rate.

Traditionally, the discount rate has been applied to policies and projects involving small changes with direct benefits and costs over less than one generation (say a few decades at most), which means that people are feeling the impact of their decisions in their own lives. However, climate change is an intergenerational policy issue, and thus, we must see ρ as a parameter capturing discrimination by date of birth. For example, applying a 2% pure time discounting rate ($\rho = 2$) gives half the ethical weight to someone born in 2008 relative to someone born in 1973. Surely, many would find this difficult to justify.

In addition, the discounting formula described above depends on the path of future growth in consumption. Climate change involves potentially very large changes and can reduce future growth in consumption, so the discount rate applied in a world with climate change will be less than that in a world without, all else being equal. Moreover, this logic can be extended so that the uncertainty around climate impacts is taken into account. For every possible scenario of future climate change, there will be a specific average discount rate, depending on the growth rate of consumption in that scenario (4). Thus, to speak of "the discount rate" is misguided.

KEY TERM

Social marginal utility is the social benefit derived from one additional unit of a product or service.

Using $\eta = 1$ implies that a given social benefit will be valued more highly by a factor of five for someone with one-fifth the resources of someone else. Some commentators have suggested that higher values should be used. Using $\eta = 2$ would mean that an extra benefit to the person who is poorer by a factor of five would have a value 25 times that to a richer person. In a transfer from the richer individual to the poorer one, how much would you be prepared to lose in the process and still regard it as a beneficial transfer? In the case of $\eta = 2$, as long as less than 96% is lost, it would be seen as beneficial; for $\eta = 1$, less than 80%. Although it is a tenable ethical position, those who argue for η as high as 2 should be advocating very strong redistribution policies.

In the case of $\eta = 3$ in Nordhaus's example,

over 99% could be lost and a transfer would still be beneficial. Does he advocate huge increases in transfers from rich to poor in the current generation?

A value of unity for η is quite commonly invoked, but higher values of ρ are sometimes used in cost-benefit analysis. Indeed, there are a number of reasons why a smaller-scale project such as a new road or railway may not be as valuable—or relevant at all—in several years time as circumstances change. However, avoiding the impacts of climate change (the value of a stable climate, human life, and ecosystems) is likely to continue to be relevant as long as the planet and its people exist.

Further, as people become richer and environmental goods become scarcer, it seems likely that, rather than fall, their value will rise very rapidly, which was an issue raised in chapter 2 of our review and has been investigated in later analyses (5). And the flow-stock nature of greenhouse gas accumulation, plus the powerful impact of climate change, will render many consequences irreversible. Thus, investing elsewhere and using the resources to compensate for any later environmental damage may be very cost-ineffective.

Many of the comments on the review have suggested that the ethical side of the modeling should be consistent with observable market behavior. As discussed by Hepburn (6), there are many reasons for thinking that market rates and other approaches that illustrate observable market behavior cannot be seen as reflections of an ethical response to the issues at hand. There is no real economic market that reveals our ethical decisions on how we should act together on environmental issues in the very long term.

Most long-term capital markets are very thin and imperfect. Choices that reflect current individual personal allocations of resource may be different from collective values and from what individuals may prefer in their capacity as citizens. Individuals will have a different attitude to risk because they have a higher probability of demise in that year than society.

Those who do not feature in the marketplace (future generations) have no say in the calculus, and those who feature in the market less prominently (the young and the poor) have less influence on the behaviors that are being observed.

The issue of ethics should be tackled directly and explicitly through discussion (7). No discussion of the appropriateness of particular value judgments can be decisive. Alternative ethical approaches should be explored: Within the narrow confines of the modeling, sensitivity analysis does this. There is also scope for further work attempting to disentangle the roles of risk aversion and inequality aversion that are conflated (via η) in this modeling. Furthermore, we should go beyond the narrow framework of social welfare functions to consider other ethical approaches, including those involving rights and sustainability.

We note briefly that Nordhaus misrepresents the Stern Review on the subject of taxes. He argues that we propose a tax of \$85 per ton of CO_2, which equates to \$312 per ton of carbon. This was our estimate of the marginal environmental cost of each extra carbon emission (the "social cost of carbon," hereafter SCC) under business as usual, with no policies to reduce emissions. To identify this with a recommended tax makes two mistakes. First, any estimate of the SCC is path-dependent. In chapter 13 of the Stern Review, we justify our proposed policy goal of stabilizing emissions between 450 and 550 ppm CO_2 equivalent. In this range, we estimate the social cost of carbon to be between \$25 per ton of CO_2 (at 450 ppm) and \$30 per ton (at 550 ppm), and so the proposed package of policies should be broadly consistent with this range. Second, in distorted and uncertain economies, any tax should be different from an SCC (8). Stabilization between 450 and 550 ppm is equivalent to reductions of around 25% to 70% in 2050. Nordhaus claims erroneously that the review suggests reductions on this scale over the next two decades. The Stern Review is also clear that prices should increase over time,

although perhaps not as sharply as Nordhaus suggests.

The ethical approach in Nordhaus's modeling helps drive the initial low level of action and the steepness of his policy ramp. As future generations have a lower weight, they are expected to shoulder the burden of greater mitigation costs. This could be a source of dynamic inconsistency, because future generations will be faced with the same challenge and, if they take the same approach, will also seek to minimize short-term costs but expect greater reductions in the future as they place a larger weight on consumption now over the effects on future generations (thus perpetuating the delay for significant reductions).

We have argued strongly for an assessment of policy on climate change to be based on a disaggregated approach to consequences—looking at different dimensions, places, and times—and a broad ethical approach. Nevertheless, our modeling sensitivity analysis demonstrates that the treatment of risk and uncertainty and the extent to which the model responds to progress in the scientific literature are of roughly similar importance in shaping damage estimates as our approach to ethics and discounting. It is these three factors that explain higher damage estimates than those in the previous literature.

Given the centrality of risk, scientific advance, and ethics, in our view, the question should really be why, with some important exemptions, did the previous literature pay inadequate attention to these issues?

There was much structural caution in our approach. We left out many risks that are likely to be important, for example, the possibility of strong disruption of carbon cycles by changes to oceans and forests. It is possible that risks and damages are higher than we estimated. But one thing is clear: however unpleasant the damages from climate change are likely to appear in the future, any disregard for the future, simply because it is in the future, will suppress action to address climate change.

References and Notes

1. W. Nordhaus, *Science* 317, 201 (2007).

2. N. Stern, *The Economics of Climate Change: The Stern Review* (Cambridge Univ. Press, Cambridge, 2006).

3. F. P. Ramsey, *Econ. J.* 38, 543 (1928).

4. In the review's modeling, the *g* in the discount rate is specific to the growth path in each of the thousands of model runs in the Monte Carlo analysis of aggregated-impact cost estimates.

5. T. Sterner, U. M. Persson, *An Even Sterner Review: Introducing Relative Prices into the Discounting Debate*, working draft, May 2007; www.hgu.gu.se/files/nationalekonomi/personal/thomas%20sterner/b88.pdf.

6. C. Hepburn, *The Economics and Ethics of Stern Discounting*, presentation at the workshop The Economics of Climate Change, 9 March 2007, University of Birmingham, Birmingham, UK; www.economics.bham.ac.uk/maddison/Cameron%20Hepburn%20Presentation.pdf.

7. N. Stern, Value judgments, welfare weights and discounting, Paper B of *After the Stern Review: Reflections and Responses*, 12 February 2007, working draft of paper published on Stern Review Web site; www.sternreview.org.uk.

8. N. Stern, The case for action to reduce the risks of climate change, Paper A of *After the Stern Review: Reflections and Responses*, working draft of paper published on Stern Review Web site, 12 February 2007; www.sternreview.org.uk.

Dealing with Climate Change

Introduction

DONALD KENNEDY

In this section, we turn from the matter of predicting the future course of climate change to the more difficult question of what we might do about it. Perhaps it is a good idea to begin with some history: What have scientists thought about what might be an appropriate limit for the ultimate concentration of greenhouse gases in the atmosphere? It may surprise readers who see this as an issue of recent origin that the first U.S. government pronunciation about the "global warming" was issued by the President's Council on Environmental Quality in the very first month of 1981. The Council, chaired by Gus Speth—now Dean of the Yale School of Forestry and Environmental Studies—predicted that a doubling of the preindustrial level of carbon dioxide (CO_2) in the atmosphere would raise the average global temperature by about 3°C, with more significant increases at the poles. It included an alarming sea level rise projection of up to 20 feet.

It is interesting how closely most of this accords with the more recent "IPCC Consensus," which puts a wider range of possibility around the temperature increase. A doubling of greenhouse gas concentrations would take us to 560 parts per million by volume (ppmv) of CO_2. The 1981 report said that a level 50% higher than the preindustrial level would be an appropriate upper limit—but that value, 420 ppmv, is one that is being approached rapidly now. Most current climate scientists are in agreement that 500 ppmv may be a value to which human societies could adapt—but they also agree that even 550 ppmv may, in the words of the Framework Convention on Climate Change from the Rio Conference, constitute "dangerous anthropogenic interference with the world's climate system." And a broad consensus holds that the "business-as-usual" scenario, in which present rates of addition of greenhouse gases continue, entails dangerous risks for human societies.

Geoengineering

The risks of climate change are so serious that some scientists believe we must intervene through geoengineering. It is a controversial topic because environmentalists often see geoengineering as a high-tech excuse for not taking the economically difficult steps toward CO_2 emissions mitigation. On the other hand, it bespeaks innovation and an encouraging commitment to addressing climate change. The different views can be seen in "Giving Climate Change a Kick," a *Science* news article in which Eli Kintisch describes a November 2007 meeting of top climate scientists sponsored by Harvard University and the University of Calgary. Their cautious consensus was that geoengineering deserves further research.

In the first research article of this section, Tom Wigley, a very accomplished climate modeler at the National Center for Atmospheric Research in Boulder, Colorado, explores some possible geoengineering approaches. One possibility, fertilizing ocean regions that lack a particular nutrient, might enhance carbon fixation—but it could also affect acidity. Wigley points out that the ejection of sulfate aerosols by the 1982 Mount Pinatubo volcano produced a two-year period of global cooling. Modeling suggests that the injection of such aerosols into the stratosphere could produce significant cooling for short periods—enough to buy time for other efforts to reduce fossil fuel consumption or sequester carbon.

Widespread interest in geoengineering is one indicator of the intensity of the current struggle to deal with global warming. The extent of research into other approaches, from building efficiency to electric or hybrid transportation vehicles to alternative fuels, also symbolizes an awakened public knowledge that humanity faces a serious crisis, and that there is no time to waste.

Carbon Capture

Capturing carbon, which may involve a number of different strategies, is one intriguing way of slowing the increase in atmospheric CO_2. This technology is covered in the next two pieces by Klaus S. Lackner, in the Department of Earth and Environmental Engineering at Columbia University, and Daniel Schrag, Director of Harvard's Center for the Environment.

Schrag, who is also Professor of Earth and Planetary Science, has been a leading investigator of historical processes in geology and oceanography, yielding insights into past climates. In his article "Preparing to Capture Carbon," he concentrates on mechanisms of capture—particularly from coal-powered electrical plants. Why coal? It is the most carbon-intensive energy source, and the world's reserves in coal dwarf those in oil or natural gas. Schrag shows how carbon can be extracted from such facilities and then stored, either underground or beneath the ocean floor. But the cost of these processes may make them unacceptable politically, at least in the short term or until new technologies are developed. Meanwhile, China and especially Russia continue to build numbers of coal-fired plants each year. In fact, Russia prefers to sell its natural gas to Europe for cash and in the interest of deploying "soft power."

Schrag mentions but does not explore the closely related concept of sequestration. Often, that term is used as a synonym for "carbon capture." But sequestration of carbon is more often thought of as its capture by plants, and then storage in woody biomass above ground, or in underground roots. In his article "A Guide to CO_2 Sequestration," Lackner argues that many sequestration methods, such as biomass sequestration and CO_2 utilization, won't make a dent in the energy budget. But he does believe carbon capture through underground injection is an important stopgap measure until alternative, clean technologies become widespread and inexpensive.

Remedies: Taxes, Trading Systems

Of course, the most direct remedy for an industrial nation like the United States is to reduce the combustion of fossil fuels. Reduction will have to be achieved by government intervention: either through taxing carbon or by issuing carbon emission permits, which can then be traded. The first of these would involve taxation at the wellhead or the mine mouth; the cost to the producers would then be passed on to industrial consumers or utilities, and thence to consumers as increases in rates. The second would create a market in which good performers in reducing emissions could sell permits to less successful ones. In a given year, the number of permits issued would be based on the CO_2 emissions reduction established as necessary, and in successive years, the number could be reduced. Most advocates of this "market regulation" system believe that the permits should be auctioned, with the government using the proceeds for managing the system or undertaking other carbon-reducing alternatives.

How to decide between these systems and evaluate each one is subject matter for the exchange on carbon trading and carbon taxes between three thought leaders of the policy discussions now under way. William Schlesinger is a long-term student of the carbon cycle, including in his studies the effects of temperature on forests in the vicinity of Duke University, where he was a faculty member until he was named President of the Cary Institute of Ecosystem Studies in Millbrook, New York. He argues that carbon-trading systems are a promising idea, but he presents taxation schemes as preferable. The crux of his argument is that whichever mitigation mechanism is proposed, it should concentrate on fossil fuel carbon and ignore forests and other sinks and sources for carbon—on the basis that the latter are too small and uncertain to be included in an international trading system.

William Chameides, an ecologist who has now become Dean of the Nicholas School of the Environment and Earth Sciences at Duke, was associated with Michael Oppenheimer as Chief Scientist at Environmental Defense before Oppenheimer moved to accept a chair in Geosciences at Princeton. Their position is that market-based tradable permit systems would constitute a more effective system for limiting CO_2 emissions. More important, Chameides and Oppenheimer would include offsetting emissions by carbon storage in forests and soils; they express confidence that mechanisms for certifying these offsets exist, and they claim that there is a potential for offsetting up to 20% of U.S. emissions at low cost.

Offsets

"Carbon offsets" have become an even larger issue in our contemporary lives in another way. Because there is a more widespread concern about global warming, there is a new interest on the part of organizations or even individuals to calculate the amount of carbon emissions for which they are responsible. Some of these offsets involve carbon-saving procedures like planting trees or protecting forests that would otherwise be burned. Some suspicions have arisen about the authenticity of some of these programs. In the Section I information about biofuels, we touched on some of the rationale for questioning whether these offsets actually do save carbon. Many projects for converting corn or sugarcane or switchgrass into ethanol as a fuel-sparing additive actually do not yield a carbon benefit significantly greater than the emissions used in growing, transporting, and refining the crop. It has been suggested that some perennial crops grown on marginal agricultural land can both sequester carbon in the soil and yield an additional carbon benefit when harvested and then either burned as biomass or refined to extract ethanol.

Waking Up

Individuals' efforts to reduce their carbon footprint through offsets or other measures point to a growing political will in the United States to combat climate change. Therefore it should not surprise the reader that the last news article in this volume explores efforts in the Senate to pass legislation that would limit carbon emissions. At the time of this writing, we are unsure whether such a bill will pass, or what it would look like in its final form. But Eli Kintisch's article is an appropriate postscript for this volume, because we stand at an important juncture. Will our newly serious awareness of the climate crisis urge the United States to assume a leading role in its solution? Will we discover or create international institutions capable of ushering in agreements among nations both rich and poor? A sustainable world economy surely depends in part on the answers.

A Combined Mitigation-Geoengineering Approach to Climate Stabilization

T. M. L. WIGLEY

Projected anthropogenic warming and increases in carbon dioxide (CO_2) concentration present a twofold threat, both from climate changes and from CO_2 directly through increasing the acidity of the oceans. Future climate change may be reduced through mitigation (reductions in greenhouse gas emissions) or through geoengineering. Most geoengineering approaches, however, do not address the problem of increasing ocean acidity. A combined mitigation-geoengineering strategy could remove this deficiency. Here we consider the deliberate injection of sulfate aerosol precursors into the stratosphere. This action could substantially offset future warming and provide additional time to reduce human dependence on fossil fuels and stabilize CO_2 concentrations cost-effectively at an acceptable level.

In the absence of policies to reduce the magnitude of future climate change, the globe is expected to warm by about 1°C to 6°C over the

KEY TERM

Geoengineering refers to the intentional large-scale manipulation of the environment to counteract anthropogenic climate change. A particular example is solar radiation management (SRM), in which part of the Sun's incoming radiation is artificially reflected away from the Earth using stratospheric aerosols of reflecting bodies outside the atmosphere.

This article first appeared in *Science* (originally published in *Science* Express on 14 September 2006, *Science* 20 October 2006: Vol. 314, no. 5798). It has been revised for this edition.

*Science*NOW Daily News, 9 November 2007

GIVING CLIMATE CHANGE A KICK
Eli Kintisch

CAMBRIDGE, MASSACHUSETTS—Top climate scientists have cautiously endorsed the need to study schemes to reverse global warming that involve directly tinkering with Earth's climate. Their position on geoengineering, which will likely be controversial, was staked out at an invitation-only meeting that ended here today. It's based on a growing concern about the rapid pace of global change and continued anthropogenic emissions of greenhouse gases.

"In this room, we've reached a remarkable consensus that there should be research on this," said climate modeler Chris Bretherton of the University of Washington, Seattle, during a morning session today. Phil Rasch, a modeler with the University Corporation for Atmospheric Research in Boulder, Colorado, underscored the point. "We're not saying that there should be geoengineering; we're saying there should be research regarding geoengineering." No formal statement was released at the meeting, which was organized by Harvard University and the University of Calgary, but few of the 50 scientists objected to the idea.

The field of geoengineering has long been big on ideas but short on respect. Some of the approaches that researchers have dreamed up include launching fleets of space-based shades to dim the sunlight hitting Earth or altering the albedo of the ocean with light-colored reflectors. Perhaps the best-known idea is to pump aerosols into the stratosphere to mimic the cooling effect of volcanoes. But there's been scant support from mainstream scientists, many of whom fear that even mentioning the g-word could derail discussion of carbon emissions cuts.

Harvard geochemist Daniel Schrag and physicist David Keith of the University of Calgary thought that geoengineering deserved a closer look (*Science*, 26 October 2007, p. 551). In an opening presentation yesterday, Schrag explained that extensive, rapid melting of arctic sea ice (*Science*NOW, 2 May 2007) and the fact that the world's 2005 and 2006 carbon emissions from fossil fuels were higher than predictions by the Intergovernmental Panel on Climate Change are forcing the hands of climate scientists. Schrag also fears that when countries are faced with the prospect of even more drastic environmental change, they will turn to geoengineering regardless of whether the consequences are known.

The degree of scientific uncertainty was clear throughout the two-day meeting. Harvard paleoclimate scientist Peter Huybers told his colleagues during one session that understanding of the world's climate may not be sufficient to properly wield geoengineering tools. "We should be humble about how much we know about the climate system," Huybers said.

Most of the discussion focused on whether to jump-start what has been an anemic research agenda with no public financing. Some participants said that they were spurred into action by a paper that appeared in *Climatic Change* last year, in which Nobelist Paul Crutzen called for geoengineering research (*Science*, 20 October 2006, p. 401).

Harvard climate researcher James Anderson told the group that the Arctic ice was "holding on by a thread" and that more carbon emissions could tip the balance. The delicacy of the system, he said "convinced me of the need for research into geoengineering," And five years ago? "I would have said it's a very inappropriate solution to the problem."

A sulfate aerosol precursor is a directly emitted chemical substance that is transformed to droplets of sulfuric acid and/or other sulfate compounds. Examples are sulfur dioxide (SO_2), emitted as a result of fossil fuel burning or volcanic activity, and dimethyl sulfide (DMS), released from the oceans.

The stratosphere (the layer of the atmosphere between approximately 6 and 30 miles above the Earth's surface) contains over 90% of the atmosphere's ozone. Ozone is a gas composed of three oxygen atoms; it absorbs much of the Sun's incoming ultraviolet radiation and prevents it from reaching the Earth.

Chlorine loading is the net stratospheric concentration of chlorine-containing, ozone-depleting substances, arising from halocarbons such as CFCs, HCFCs, and HFCs that all contain chlorine. Chlorine loading is directly related to the amount of ozone depletion.

21st century (1, 2). Estimated CO_2 concentrations in 2100 lie in the range from 540 to 970 parts per million (ppm), which is sufficient to cause substantial increases in ocean acidity (3–6). Mitigation directed toward stabilizing CO_2 concentrations (7) addresses both prob-

lems but presents considerable economic and technological challenges (8, 9). Geoengineering (10–17) could help reduce the future extent of climate change due to warming but does not address the problem of ocean acidity. Mitigation is therefore necessary, but geoengineering could provide additional time to address the economic and technological challenges faced by a mitigation-only approach.

The geoengineering strategy examined here is the injection of aerosol or aerosol precursors [such as sulfur dioxide (SO_2)] into the stratosphere to provide a negative forcing of the climate system and consequently offset part of the positive forcing due to increasing greenhouse gas concentrations (18). Volcanic eruptions provide ideal experiments that can be used to assess the effects of large anthropogenic emissions of SO_2 on stratospheric aerosols and climate. We know, for example, that the Mount Pinatubo eruption [June 1991 (19, 20)] caused detectable short-term cooling (19–21) but did not seriously disrupt the climate system. Deliberately adding aerosols or aerosol precursors to the stratosphere, so the loading is similar to the maximum loading from the Mount Pinatubo eruption, should therefore present minimal climate risks.

Increased sulfate aerosol loading of the stratosphere may present other risks, such as through its influence on stratospheric ozone. This particular risk, however, is likely to be small. The effect of sulfate aerosols depends on the chlorine loading (22–24). With current elevated chlorine loadings, ozone loss would be increased. This result would delay the recovery of stratospheric ozone slightly but only until anthropogenic chlorine loadings returned to levels of the 1980s (which are expected to be reached by the late 2040s).

Figure 1 shows the projected effect of multiple sequential eruptions of Mount Pinatubo every year, every two years, and every four years. The Pinatubo eruption–associated forcing that was used had a peak annual mean value of −2.97 watts per square meter (W/m^2) (20, 21). The climate simulations were carried out using

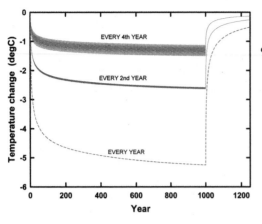

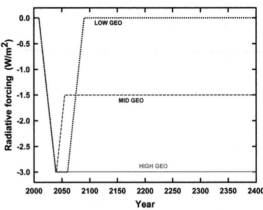

FIGURE. 1. (left) Global mean temperature response to multiple volcanic eruptions. The standard eruption used was that of Mount Pinatubo [forcing data from Ammann et al. (20, 21)], and eruptions were assumed to occur every four years (top curve), every two years (middle curve), or every year (bottom curve). The results shown are annual mean values plotted year by year. In the two- and (especially) four-year cases, the forcing varies considerably from year to year, leading to noticeable interannual temperature variations. These appear as bands of values because the abscissa scale in the graph is insufficient to resolve these rapid variations. A climate sensitivity of 3°C equilibrium warming for $2 \times CO_2$ is assumed.

FIGURE 2. (right) Radiative forcing scenarios for the three geoengineering options considered. The HIGH GEO option corresponds approximately to the steady-state forcing that would result from eruptions of Mount Pinatubo every two years.

an upwelling-diffusion energy-balance model [Model for the Assessment of Greenhouse gas–Induced Climate Change (MAGICC) (2, 25, 26): a simple type of climate model that considers the climate effects of changes in the balance between incoming and outgoing energy to the Earth-ocean-atmosphere system] with a chosen climate sensitivity of 3°C equilibrium warming for a CO_2 doubling ($2 \times CO_2$). Figure 1 suggests that a sustained stratospheric forcing of about -3 W/m^2 (the average asymptotic forcing for the biennial eruption case) would be sufficient to offset much of the anthropogenic warming expected over the next century. Figure 1 also shows how rapidly the aerosol-induced cooling disappears once the injection of material into

the stratosphere stops, as might become necessary should unexpected environmental damages arise.

Three cases are considered to illustrate possible options for the timing and duration of aerosol injections. In each case, the loading of the stratosphere begins in 2010 and increases linearly to -3 W/m^2 over 30 years. The options depart from each other after this date (Fig. 2). These geoengineering options are complemented by three future CO_2 emissions scenarios: a central "no climate policy" scenario from the Special Report on Emissions Scenarios (SRES) (27) data set, namely the A1B scenario; an ambitious scenario known as WRE450 (7) in which CO_2 concentration stabilizes at 450 ppm (the present level is about 380 ppm); and an overshoot scenario in which CO_2 concentration rises to 530 ppm in 2080 before declining to 450 ppm. [Because an atmospheric CO_2 concentration of 450 ppm "produces both calcite and aragonite undersaturation in most of the deep ocean" (4), a level even less than this may ultimately be desirable.]

CO_2 concentrations and corresponding fos-

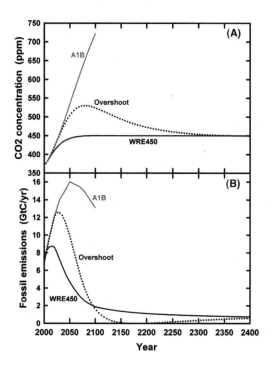

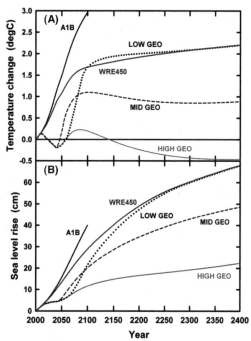

FIGURE 3. (left) (A) CO_2 concentration projections used in the analysis together with (B) corresponding fossil fuel emissions. The overshoot scenario was used in conjunction with the three geoengineering options shown in Fig. 2. A climate sensitivity of 3°C equilibrium warming for $2 \times CO_2$ is assumed. CO_2 emissions results depend on the climate sensitivity because of climate feedbacks on the carbon cycle. GtC, gigatons of carbon.

FIGURE 4. (right) Global mean temperature (A) and sea level (B) changes for the A1B scenario, the WRE450 scenario, and three scenarios combining both mitigation and geoengineering. The latter cases use the overshoot scenario (Fig. 3) and the three increasingly strong geoengineering options (Fig. 2). A climate sensitivity of 3°C equilibrium warming for $2 \times CO_2$ is assumed.

sil fuel emissions for these three CO_2 scenarios are shown in Figure 3. Emissions for the stabilization cases were calculated with the use of an inverse version of MAGICC, which accounted for climate feedbacks on the carbon cycle. The WRE450 scenario is an archetypal mitigation-only case, stabilizing at a level that many researchers believe would avoid "dangerous anthropogenic interference" with the climate system (28). The overshoot scenario is introduced here to be considered in conjunction with the three geoengineering options. It allows much more CO_2 emission and a much slower departure from the A1B no-policy scenario baseline. Although the rate of decline

of emissions in the mid-to-late-21st century is more rapid in the overshoot scenario than in WRE450, these reductions begin 15 to 20 years later in the former scenario, allowing additional time both to phase out existing CO_2-emitting fossil fuel energy technologies and to develop and deploy energy sources that have net-zero CO_2 emissions (7–9).

Figure 4 shows global mean temperature and sea level projections for the no-policy scenario (A1B), the mitigation-only scenario (WRE450), and the overshoot CO_2 scenario, combined with the three alternative geoengineering options (HIGH GEO, MID GEO, and LOW GEO). For the decades immediately after 2010, changes in aerosol forcing in all three GEO options occur more rapidly than forcing changes for the CO_2 scenarios, so the net effect is cooling. After 2040, the HIGH GEO–associated cooling tends to balance the warming from the overshoot CO_2 stabilization scenario, eventually leading to a slight cooling that would bring global mean temperatures back to near their preindustrial level. The MID and LOW GEO options lead to temperatures stabilizing at approximately 1°C and 2°C above temperatures in 2000 (29). After 2100, the LOW GEO option (where injection into the stratosphere is decreased to zero by 2090) closely matches the WRE450 mitigation-only scenario but requires less stringent emissions reductions.

The sea level results (Fig. 4B), derived from models used in the Third Assessment Report of the Intergovernmental Panel on Climate Change (30, 31), show the much larger inertia of this part of the climate system. The LOW GEO option and WRE450 scenario again are similar, with neither tending toward stabilization. Even the HIGH GEO option shows a continuing (but slow) rise in sea level toward the end of the study period, but the rate of rise is small, even relative to changes observed over the 20th century (30, 32).

A combined mitigation-geoengineering approach to climate stabilization has a number of advantages over either alternative used separately. A relatively modest geoengineering investment (33, 34) corresponding to the present LOW GEO option could reduce the economic and technological burden on mitigation substantially, by deferring the need for immediate or near-future cuts in CO_2 emissions. More ambitious geoengineering, when combined with mitigation, could even lead to the stabilization of global mean temperature at near present levels and reduce future sea level rise to a rate much less than that observed over the 20th century: aspects of future change that are virtually impossible to achieve through mitigation alone.

As a guide to the amount of SO_2 required, the eruption of Mount Pinatubo injected about 10

teragrams of sulfur (TgS) into the stratosphere (35, 36), and the analysis here suggests that an annual flux of half that amount would have a substantial influence. Smaller-diameter aerosols would have longer lifetimes and require still smaller injection rates (15). Five TgS/year is only about 7% of current SO_2 emissions from fossil fuel combustion (37, 38). Further analysis is required to assess (i) the technological feasibility of the suggested injections of SO_2 [or of more radiatively efficient material (34)] into the stratosphere, (ii) the economic costs of this option relative to the reduced costs of mitigation that an overshoot CO_2–stabilization pathway would allow, and (iii) the detailed effects of the proposed SO_2 injections and CO_2 concentration changes on climate [compare with (39)] and stratospheric chemistry (40).

References and Notes

1. U. Cubasch, G. A. Meehl, in *Climate Change 2001: The Scientific Basis*, J. T. Houghton *et al.*, Eds. (Cambridge Univ. Press, Cambridge, 2001), pp. 525–582.

2. T. M. L. Wigley, S. C. B. Raper, *Science* 293, 451 (2001).

3. K. Caldeira, M. E. Wickett, *Nature* 425, 365 (2003).

4. K. Caldeira, M. E. Wickett, *J. Geophys. Res.* 110, 10.1029/2004JC002671 (2005).

5. J. C. Orr *et al.*, *Nature* 437, 681 (2005).

6. A. J. Andersson, F. T. Mackenzie, A. Lerman, *Global Biogeochem. Cycles* 20, 10.1029/2005GB002506 (2006).

7. T. M. L. Wigley, R. Richels, J. A. Edmonds, *Nature* 379, 240 (1996).

8. M. I. Hoffert *et al.*, *Nature* 395, 881 (1998).

9. M. I. Hoffert *et al.*, *Science* 298, 981 (2002).

10. Geoengineering refers to engineering aimed at counteracting the undesired side effects of other human activities (11). Two classes may be distinguished: climate engineering (12–15) and carbon cycle engineering [for example, through modifying ocean alkalinity (16, 17)]. Here, the word *geoengineering* refers specifically to climate engineering (i.e., direct and intentional management of the climate system).

11. D. W. Keith, *Annu. Rev. Energy Environ.* 25, 245 (2000).

12. S. H. Schneider, *Nature* 409, 417 (2001).

13. National Academy of Sciences, *Policy Implications of Greenhouse Warming: Mitigation, Adaptation*

and the Science Base (National Academy Press, Washington, DC, 1992), pp. 433–464.

14. B. P. Flannery et al., in Engineering Response to Global Climate Change: Planning a Research and Development Agenda, R. G. Watts, Ed. (Lewis Publishers, Boca Raton, FL, 1997), pp. 379–427.

15. P. J. Crutzen, Clim. Change 76, 10.1007/s10584-006-9101-y (2006).

16. H. S. Kheshgi, Energy 20, 915 (1995).

17. K. S. Lackner, Annu. Rev. Energy Environ. 27, 193 (2002).

18. M. I. Budyko, Climate Changes (Engl. transl.) (American Geophysical Union, Washington, DC, 1977).

19. A. Robock, Rev. Geophys. 38, 191 (2000).

20. C. M. Ammann, G. A. Meehl, W. M. Washington, C. S. Zender, Geophys. Res. Lett. 30, 10.1029/2003GL016875 (2003).

21. T. M. L. Wigley, C. M. Ammann, B. D. Santer, S. C. B. Raper, J. Geophys. Res. 110, 10.1020/2004JD005557 (2005).

22. X. Tie, G. Brasseur, Geophys. Res. Lett. 22, 3035 (1995).

23. S. Solomon et al., J. Geophys. Res. 101, 6713 (1996).

24. R. W. Portmann et al., J. Geophys. Res. 101, 22991 (1996).

25. MAGICC is a suitable tool because it reproduces the results from more complex coupled ocean-atmosphere general circulation models for both volcanic time-scale (21) and long-term (26) forcing.

26. S. C. B. Raper, J. M. Gregory, T. J. Osborn, Clim. Dyn. 17, 601 (2001); http://www.sciencemag.org/cgi/external_ref?access_num=10.1007/PL00007931&link_type=DOI.

27. N. Nakićenović, R. Swart, Eds., Special Report on Emissions Scenarios (Cambridge Univ. Press, Cambridge, 2000).

28. Avoiding dangerous anthropogenic interference with the climate system is one of the primary guidelines for climate policy espoused in Article 2 of the United Nations Framework Convention on Climate Change.

29. In all options, there is a residual warming tendency arising from the emissions of non-CO_2 gases (such as CH_4, N_2O, halocarbons, and tropospheric aerosols). Emissions from these sources are assumed to follow the A1B scenario to 2100 and then remain constant, leading to a slow but long-term increase in radiative forcing.

30. J. A. Church, J. M. Gregory, in Climate Change 2001: The Scientific Basis, J. T. Houghton et al., Eds. (Cambridge Univ. Press, Cambridge, 2001), pp. 639–694.

31. T. M. L. Wigley, S. C. B. Raper, Geophys. Res. Lett. 32, 10.1029/2004GL021238 (2005).

32. A. Cazenave, R. S. Nerem, Rev. Geophys. 42, 10.1029/2003RG000139 (2004).

33. Teller et al. (34) point out that sulfate aerosols are "grossly nonoptimized" as scatterers of short-wave radiation and that metallic or resonant scatterers offer large mass savings. Although emplacement costs for such scatterers would be higher, they estimate net costs (for metals) to be "as much as five times less" than for sulfate aerosols.

34. E. Teller, R. Hyde, L. Wood, Active Climate Stabilization: Practical Physics-Based Approaches to Prevention of Climate Change, Preprint UCRL-JC-148012 (Lawrence Livermore National Laboratory, Livermore, CA, 2002); www.llnl.gov/global-warm/148012.pdf.

35. G. J. S. Bluth, S. D. Doiron, C. C. Schnetzler, A. J. Krueger, L. S. Walter, Geophys. Res. Lett. 19, 151 (1992).

36. S. Guo, G. J. S. Bluth, W. I. Rose, I. M. Watson, A. J. Prata, Geochem. Geophys. Geosys. 5, 10.1029/2003GC000654 (2004).

37. S. J. Smith, H. Pitcher, T. M. L. Wigley, Global Planet. Change 29, 99 (2001).

38. At steady-state conditions, an injection of 5 TgS/year into the stratosphere would increase the flux into the troposphere by the same amount, with larger fluxes in high latitudes. Given current emissions, the consequences of this extra flux are likely to be minor, but this aspect warrants further attention.

39. B. Govindasamy, K. Caldeira, Geophys. Res. Lett. 17, 2151 (2000).

40. The National Center for Atmospheric Research is supported by NSF.

Preparing to Capture Carbon

DANIEL P. SCHRAG

arbon sequestration from large sources of fossil fuel combustion, particularly coal, is an essential component of any serious plan to avoid catastrophic impacts of human-induced climate change. Scientific and economic challenges still exist, but none are serious enough to suggest that carbon capture and storage will not work at the scale required to offset trillions of tons of carbon dioxide (CO_2) emissions over the next century. The challenge is whether the technology will be ready when society decides that it is time to get going.

Strategies to lower CO_2 emissions to mitigate climate change come in three flavors: reducing the amount of energy the world uses, either through more efficient technology or through changes in lifestyles and behaviors; expand-

ing the use of energy sources that do not add CO_2 to the atmosphere; and capturing the CO_2 from places where we do use fossil fuels and then storing it in geologic repositories, a process known as carbon sequestration. A survey of energy options makes clear that none of these is a silver bullet. The world's energy system is too immense, the thirst for more and more energy around the world too deep, and our dependence on fossil fuels too strong. All three strategies are essential, but the one we are furthest from realizing is carbon sequestration.

The crucial need for carbon sequestration can be explained with one word: coal. Coal produces the most CO_2 per unit energy of all fossil fuels, nearly twice as much as natural gas. And unlike petroleum and natural gas, which are predicted to decline in total production well before the middle of the century, there is enough coal to last for centuries, at least at current rates of use, and that makes it cheap relative to almost every other source of energy (Table 1). Today, coal

This article first appeared in *Science* (9 February 2007: Vol. 315, no. 5813). It has been revised for this edition.

TABLE 1. Carbon content in gigatons (Gt) of fossil fuel proven reserves and annual production (2005) (6).

Country/region	Coal Reserves	Coal Production	Petroleum Reserves	Petroleum Production	Natural gas Reserves	Natural gas Production
United States	184.0	0.64	3.6	0.30	3.0	0.29
Russia	117.1	0.15	9.0	0.42	26.2	0.33
China	85.4	1.24	1.9	0.16	1.3	0.03
India	69.0	0.22	0.7	0.03	0.6	0.02
Australia	58.6	0.23	0.5	0.02	1.4	0.02
Middle East	0.3	0.00	90.2	1.11	39.4	0.16
Total world	678.2	3.23	145.8	3.59	98.4	1.51

and petroleum each account for roughly 40% of global CO_2 emissions. But by the end of the century, coal could account for more than 80%. Even with huge improvements in efficiency and phenomenal rates of growth in nuclear, solar, wind, and biomass energy sources, the world will still rely heavily on coal, especially the five countries that hold 75% of world reserves [see (6)]: the United States, Russia, China, India, and Australia (1).

As a technological strategy, carbon sequestration need not apply only to coal plants; indeed, any point source of CO_2 can be sequestered, including biomass combustion, which would result in negative emissions. Carbon sequestration also refers to enhanced biological uptake through reforestation or fertilization of marine phytoplankton. But the potential to enhance biological uptake of carbon pales in comparison to coal emissions, ever more so as India, China, and the United States expand their stock of coal-fired power plants. So developing and deploying the technologies to use coal without releasing CO_2 to the atmosphere may well be the most critical challenge we face, at least for the next 100 years, until the possibility of an affordable and completely non–fossil energy system can be realized.

If carbon sequestration from coal combustion is essential to mitigate the worst impacts of global warming, what stands in the way of its broad implementation, both in the United States and around the world? With limited coal reserves, countries in the European Union have chosen to emphasize climate mitigation strategies that focus on energy efficiency, renewable sources, and nuclear power. Of the major coal producers, Russia, China, and India have been unwilling to sacrifice short-term economic growth, although Chinese coal gasification efforts, which many see as a step toward sequestration capacity, are more advanced than current U.S. policies. In the United States, there are scientific and economic questions that must be answered before large-scale deployment can be achieved. But none of these is critical enough to suggest that carbon sequestration cannot be done. The real obstacle is political will, which may require more dramatic public reaction to climate change impacts before carbon sequestration becomes a requirement for burning coal. In the meantime, there are critical steps that can be taken that will prepare us for the moment when that political will finally arrives.

The scientific questions about carbon sequestration are primarily associated with concerns about the reliability of storage of vast quantities of CO_2 in underground repositories. Will the CO_2 escape? The good news is that the reservoirs do not have to store CO_2 forever, just long enough to allow the natural carbon cycle to reduce the atmospheric CO_2 to near preindustrial levels. The ocean contains 50 times as much carbon as the atmosphere, mostly in the deep ocean, which has yet to equilibrate with the CO_2 from fossil fuel combustion. Over the time scale of mixing of the deep ocean, roughly 1000 to 2000 years, natural uptake of CO_2 by the

The dissolution of marine carbonates, including biogenic magnesian calcites (from coralline algae), aragonite (from corals and pteropods), and calcite (from coccolithophorids and foraminifera), neutralizes anthropogenic CO_2 and adds to the total alkalinity of the oceans.

ocean, combined with dissolution of **marine carbonats**, will absorb 90% of the carbon released by human activities. As long as the geologic storage of CO_2 can prevent substantial leakage over the next few millennia, the carbon cycle can handle it.

Our current understanding of CO_2 injection in sedimentary reservoirs on land suggests that leakage rates are likely to be very low (2). Despite many years of experience with injection of CO_2 for enhanced oil recovery, few studies have accurately measured the leakage rates over time intervals long enough to be certain that the CO_2 will stay put even for the next few centuries. In most of the geological settings under consideration, such as deep saline aquifers or old oil and gas fields, CO_2 exists as a **supercritical fluid** with roughly half the density of water. CO_2 is trapped by low-permeability cap rocks and by capillary forces, but it can escape if sedimentary formations are compromised by fractures, faults, or old drill holes. The handful of test sites around the world each inject roughly 1 million tons of CO_2 per year, a tiny amount compared with the need for as much as 10 billion tons per year by the middle of the century. An important question is whether leakage rates will rise as more and more CO_2 is injected and the reservoirs fill. It seems likely that many geological settings will provide adequate storage, but the data to demonstrate this do not yet exist. A more expan-

sive program aimed at monitoring underground CO_2 injections in a wide variety of geological settings is essential.

A recent proposal identified a leak-proof approach to storage by injecting CO_2 in sediment below the sea floor (3), which avoids the hazards of direct ocean injection, including impacts on ocean ecology. In this case, CO_2 would stay separate from the ocean because it would exist in the sediment at high pressure and low temperature as a dense liquid or combined with pore fluid as solid hydrate. Despite higher possible costs, this approach may be important for coastal locations, which are far from appropriate sedimentary basins, and it may also avoid expensive monitoring efforts if leakage from terrestrial settings is found to be a major problem.

In terms of capacity, the requirements are indeed vast. Conservative estimates of reservoir needs over the century are more than 1 trillion

A **supercritical fluid** is any substance at a temperature and pressure above its thermodynamic critical point, which represents the highest combination of temperature and pressure at which the substance can stably exist as a vapor and liquid simultaneously. A supercritical fluid can diffuse through solids in the manner of a gas, and it can dissolve materials in the manner of a liquid. Supercritical fluids can be substituted for organic solvents in a range of industrial and laboratory processes. Carbon dioxide and water are the most commonly used supercritical fluids, used for decaffeination and power generation, respectively.

tons of CO_2 and might exceed twice that much. This far exceeds the capacity of oil and gas fields, which will be among the first targets for sequestration projects because of additional revenues from enhanced oil recovery. Fortunately, the capacity of deep saline aquifers and deep-sea sediments is more than enough to handle centuries of world coal emissions (3, 4). This means that the locations first used to store CO_2 underground may not be the ones used by the middle of the century as sequestration efforts expand. It suggests that a broad research program must be encouraged that focuses not just on what will be done in the next few decades, but also on approaches that will be needed at the scale when all coal emissions will be captured.

Additional questions surround the more expensive part of carbon sequestration, the capture of CO_2 from a coal-fired power plant. Conventional pulverized-coal plants burn coal in air, producing a low-pressure effluent composed of 80% nitrogen, 12% CO_2, and 8% water. CO_2 can be **scrubbed** from the nitrogen, using amine liquids or other chemicals, and then extracted and compressed for injection into storage locations. This uses energy, roughly 30% of the energy from the coal combustion in the first place (4), and may raise the generating cost of electricity from coal by 50% (5), although these estimates are uncertain given that there is not yet a coal plant that practices carbon sequestration. Pulverized-coal plants can also be retrofitted to allow for combustion of coal in pure oxygen, although the separation of oxygen from air is similarly energy-intensive, and the modifications to the plant would be substantial and likely just as costly (4).

Gasification of coal, which involves heating and adding pure oxygen to make a mixture of carbon monoxide and hydrogen, can be used either for synthesis of liquid fuels or for electricity. These plants can be designed to produce concentrated streams of pressurized CO_2, often referred to as "capture ready," although this also comes at a high cost. Much attention has been given to coal

gasification as a means for promoting carbon sequestration, because studies suggest that the costs are lower than retrofitting an existing pulverized-coal plant (4). However, experience with gasification plants is limited; there are only two such plants in the United States, and neither is capture ready. More encouragement of coal gasification technology is important to discover whether the promises of lower sequestration costs can be realized. But regardless of the emphasis on such advanced coal plants, the world's existing arsenal of pulverized-coal plants (excluding the 150 new pulverized-coal plants that are currently in the permitting process in the United States) produce roughly 8 billion tons of CO_2 per year, more than any responsible climate change policy can accommodate. Thus, the investment in advanced coal-gasification plants must be matched by an effort to optimize our ability to capture the CO_2 from existing pulverized-coal plants.

Compared with the cost of most renewable

The United States and the world need carbon sequestration— not right now, but soon and at an enormous scale.

Who will be responsible
if CO_2 leaks?

energy sources, increasing the cost of electricity from coal by 50% to add sequestration seems like a bargain. When one includes the distribution and delivery charges, electric bills of most consumers would rise only 20% or so. So why is this not a higher priority in climate change legislation? Most legal approaches to climate change mitigation have focused on market mechanisms, primarily cap and trade programs. A problem is that the cap in Europe and any of the caps under discussion in the U.S. Congress yield a price on carbon that is well below the cost of capture and storage. Even if the cap were adjusted, power companies might hesitate to invest in the infrastructure required for sequestration, because of volatility in the price of carbon. Thus, it seems that another mechanism is required, at least to get carbon sequestration projects started.

And there are many other questions. Who will certify a storage site as appropriate? How will the capacity be determined? Who will be responsible if CO_2 leaks? How will we safeguard against cheating? It is clear that governments need to play some role in CO_2 storage, just as they do in other forms of waste disposal, but the exact details of a policy are unlikely to be decided in the near future, long before carbon sequestration becomes normal practice. But the uncertainty about these and other issues

contributes to a general cloudiness that discourages industry from making investments toward sequestration efforts.

Despite these obstacles, a variety of carbon sequestration activities are proceeding. Regional partnerships have been established in the United States, supported by the U.S. Department of Energy (DOE), to study the possibilities for sequestration around the country. In 2003, President Bush announced a commitment to FutureGen, a DOE project to build a zero-emission coal gasification plant that would capture and store all the CO_2 it produced. FutureGen is an exciting step forward, but a single coal gasification plant that demonstrates carbon sequestration is unlikely to convince the world that carbon sequestration is the right strategy to reduce CO_2 emissions. Moreover, a power plant operated by the government may fail to convince power companies that the costs of sequestration are well determined.

Luckily, FutureGen has competitors. British Petroleum (BP), in cooperation with General Electric, plans to build two electricity-generating plants, one in Scotland and one in California, that would sequester CO_2 with enhanced oil recovery. Xcel Energy has also made a commitment to build a coal gasification plant with sequestration. And more projects may soon be announced, as companies begin to view legislation controlling CO_2 emissions as a political inevitability.

Given the current questions about sequestration technology, the current economic realities that make it unlikely that many companies will invest in sequestration over a sustained period, and the political realities that make it unlikely we will see in the next few years a price on carbon high enough to force sequestration from coal, what can government do to make sure that carbon sequestration is ready when we need it? Whatever the path, it is time to get going, not just with small test projects, but with full-scale industrial experiments. The announcements by BP and Xcel Energy are encouraging because the world needs many such sequestration projects operating at different locations, with a handful of capture strategies and a wider variety of geological settings for storage. The U.S. government can encourage these efforts, and sponsor additional ones, making sure that there are 10 to 20 large sequestration projects operating for the next decade so that any problems that do arise with capture or storage can be identified. By creating a competitive bidding process for long-term sequestration contracts, the United States can ensure that the most cost-efficient strategies will be used while testing a variety of capture and storage options, including retrofitting older pulverized-coal plants. The United States and the world need carbon sequestration—not right now, but soon and at an enormous scale. Our challenge today is to ensure that the technology is ready when serious political action on climate change is finally taken (7).

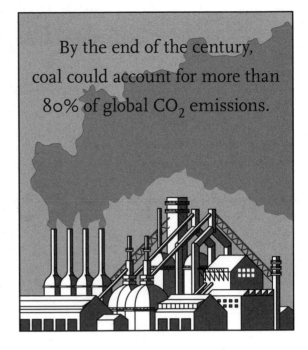

By the end of the century, coal could account for more than 80% of global CO_2 emissions.

References and Notes

1. British Petroleum, *BP Statistical Review of World Energy* (2006); www.bp.com/productlanding.do?categoryId=6842&contentId=7021390.

2. J. Bradshaw, C. Boreham, F. La Pedalina, *Storage Retention Time of CO_2 in Sedimentary Basins: Examples from Petroleum Systems*, Report of the Greenhouse Gas Technologies Cooperative Research Centre, Canberra (2004); www.co2crc.com.au/PUBFILES/STOR0405/GHGT7_Bradshaw_Boreham_LaPedalina.pdf.

3. K. Z. House, D. P. Schrag, C. F. Harvey, K. S. Lackner, *Proc. Natl. Acad. Sci. USA.* 103, 12291 (2006).

4. *IPCC Special Report on Carbon Dioxide Capture and Storage* (Intergovernmental Panel on Climate Change, 2005); www.ipcc.ch/activity/srccs/index.htm.

5. S. Anderson, R. Newell, *Annu. Rev. Environ. Resour.* 29, 109 (2004).

6. Data on proven reserves and production are from (1). Carbon content was calculated assuming 25.4 tons (1 ton = 10^3 kg) of carbon per terajoule (TJ) of coal, 19.9 tons of carbon per TJ of petroleum, and 14.4 tons of carbon per TJ of natural gas. Differences between anthracite, bituminous, lignite, and subbituminous coal were not included.

7. The author benefited from discussions with K. House and J. Holdren.

A Guide to CO$_2$ Sequestration

KLAUS S. LACKNER

Climate change concerns may soon force drastic reductions in carbon dioxide (CO$_2$) emissions. In response to this challenge, it may prove necessary to render fossil fuels environmentally acceptable by capturing and sequestering CO$_2$ until other inexpensive, clean, and plentiful technologies are available.

Today's fossil fuel resources exceed 5000 gigatons of carbon (GtC) (1), compared with world consumption of 6 GtC/year, assuring ample transition time. However, by 2050, the goal of stabilizing the atmospheric CO$_2$ concentration while maintaining healthy economic growth may require "carbon-neutral" energy in excess of today's total energy consumption (2). Lowering world CO$_2$ emissions to 2 GtC/year

would shrink the per-capita emission allowance of a projected world population of 10 billion people to 3% of today's per-capita emission in the United States.

If sequestration is to achieve this goal, it must operate on a multi-terawatt scale while sequestering almost all produced CO$_2$. It must also be safe, environmentally acceptable, and stable. For small stored quantities, storage time requirements can be minimal (3). But as storage space fills up, lifetime constraints due to aggregate **leakage emissions** would tighten, until storage times for the entire carbon stock

KEY TERM

Leakage emissions are the contribution to the world's total CO$_2$ emissions from CO$_2$ that is escaping from storage sites.

This article first appeared in *Science* (13 June 2003: Vol. 300, no. 5626). It has been revised for this edition.

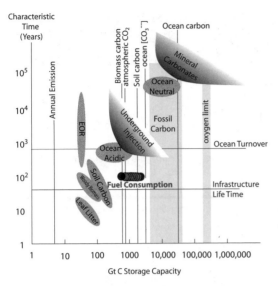

FIGURE 1. Estimated storage capacities and times for various sequestration methods. The "fossil carbon" range includes at its upper end methane hydrates from the ocean floor. The "oxygen limit" is the amount of fossil carbon that would use up all oxygen available in air for its combustion. Carbon consumption for the 21st century ranges from 600 Gt (current consumption held constant) to 2400 Gt. "Ocean acidic" and "ocean neutral" are the ocean's uptake capacities for carbonic acid and neutralized carbonic acid, respectively. The upper limits of capacity or lifetime for underground injection and mineral carbonates are not well constrained. (EOR=enhanced oil recovery.)

would reach tens of thousands of years. If carbon emissions are reduced mainly through sequestration, then total carbon storage in the 21st century will likely exceed 600 GtC. Because leaking just 2 GtC/year could force future generations into carbon restriction or recapture programs, even initial storage times should be measured in centuries.

Storage time and capacity constraints render many sequestration methods—such as **biomass sequestration** and CO_2 utilization—irrelevant or marginal for balancing the carbon budget of the 21st century. Even the ocean's capacity for absorbing carbonic acid is limited relative

to fossil carbon resources (4). Moreover, with natural ocean turnover times of centuries, storage times are comparatively short. Generally, sequestration in environmentally active carbon pools (such as the oceans) seems ill advised because it may trade one environmental problem for another.

Underground injection is probably the easiest route to sequestration. It is a proven technology suitable for large-scale sequestration (5). Injecting CO_2 into reservoirs in which it displaces and mobilizes oil or gas could create economic gains that partly offset sequestration costs. In Texas, this approach already consumes about 20 million tons/year of CO_2 at a price of $10 to $15 per ton of CO_2. However, this is not sequestration, because most of the CO_2 is extracted from underground wells.

Oil and gas sites have limited capacity (see Fig. 1). Once they fill up, saline aquifers may be used, as demonstrated under the North Sea where the Norwegian energy company Statoil has sequestered CO_2 removed from natural gas (6, 7). Ubiquitous saline reservoirs imply huge storage capacities. However, because of uncertainties in storage lifetimes, seismic instability, and potential migration of buoyant CO_2, long-term integrity must be established for each site.

*Science*NOW Daily News, 28 June 2006

A POSSIBLE SNAG IN BURYING CO$_2$

Richard A. Kerr

Scientists testing the deep geologic disposal of the greenhouse gas carbon dioxide (CO$_2$) are finding that it's staying where they put it, but it's chewing up minerals. The reactions have produced a nasty mix of metals and organic substances in a layer of sandstone 1550 m down, researchers report this week in *Geology*. At the same time, the CO$_2$ is dissolving a surprising amount of the mineral that helps keep the gas where it's put. Nothing is leaking out so far, but the phenomenon will need a closer look before such carbon sequestration can help ameliorate the greenhouse problem, say the researchers.

Drillers often inject CO$_2$ into the ground to drive more oil out, but researchers conducting the U.S. Department of Energy–sponsored Frio Brine Pilot Experiment northeast of Houston, Texas, pumped 1600 tons of CO$_2$ into the Frio Formation to see where the gas went and what it did. "We're the first looking in this huge detail so that we can see what's going on," says geochemist and lead study author Yousif Kharaka of the U.S. Geological Survey in Menlo Park, California. He and colleagues found that the CO$_2$ dropped the pH of the formation's brine from a near-neutral 6.5 to 3.0, about as acid as vinegar. That change in turn dissolved "many, many minerals," says Kharaka, releasing metals such as iron and manganese. Organic matter entered solution as well, and relatively large amounts of carbonate minerals dissolved.

The loss of carbonates worries Kharaka particularly. These naturally occurring chemicals seal pores and fractures in the rock that, if opened, could release CO$_2$ as well as fouled brine into overlying aquifers that supply drinking and irrigation water. Perhaps more troubling, says Kharaka, is that the acid mix could attack carbonate in the cement seals plugging abandoned oil or gas wells, 2.5 million of which pepper the United States. The lesson is that "whatever we do [with CO$_2$], there are environmental implications that we have to deal with," he says.

Geologist Julio Friedmann of Lawrence Livermore National Laboratory is less concerned about corrosion eating away the seals on a sequestration site. "The crust of Earth is well configured to contain CO$_2$," he says. He points to 80 U.S. oil fields injected with CO$_2$ for up to 30 years. "We've seen no catastrophic failures." Nevertheless, the Frio results do "suggest an aspect of risk we hadn't considered before," says Friedmann. There is a "new potential risk should CO$_2$ leak into shallow aquifers."

A more expensive but safer and more permanent method of CO$_2$ disposal is the neutralization of carbonic acid to form carbonates or bicarbonates (4). Neutralization-based sequestration accelerates natural weathering processes that are exothermic and thermodynamically favored, and it results in stable products that are common in nature. Mineral deposits larger than fossil resources ensure essentially unlimited supplies of base ions (mainly magnesium and calcium, but also sodium and potassium).

The least expensive way to neutralize CO$_2$ may be its injection into alkaline mineral strata. CO$_2$ would gradually dissolve into the pore water. Because it is acidic, it would leach mineral base from the rock, resulting in carbonates or bicarbonates that eliminate all concerns over long-term leakage. Neutralizing carbonic acid with carbonates as a base would create aqueous bicarbonate solutions. Unless injected underground, they would likely find their way into the ocean, which fortunately could accept far larger amounts of bicarbonates than of carbonic acid.

A better option than forming water-soluble bicarbonates would be the formation of insoluble carbonates that could be stored at the location of the mineral base, confining environmental impact to a specific site. To this end, serpentine or olivine rocks rich in magnesium silicates can be mined, crushed, milled, and reacted with CO_2. Estimated mining and mineral preparation costs of less than $10 per ton of CO_2 seem acceptable, adding 0.5 cent to 1 cent to a kilowatt-hour of electricity.

Improved methods for accelerating carbonation are, however, still needed. The current best approach—carbonation of heat-treated peridotite or serpentine in an aqueous reaction—is too costly. Elimination of the energy-intensive heat treatment could render the process economically and energetically feasible. Above-ground mineral sequestration has the capacity of binding all CO_2 that could ever be generated and limiting the environmental impact, including terrain changes, to relatively confined areas.

Most sequestration methods require concentrated CO_2, which is best captured at large plants that generate clean, carbon-free energy carriers such as electricity and hydrogen. Retrofitting existing plants appears too expensive; new plants designed for CO_2 capture are more promising (8). Complete CO_2 capture opens the door to radically new power plant designs that eliminate all flue gas emissions, not only CO_2. Oxygen-blown gasification or combustion could approach this goal today. More advanced designs could even remove the efficiency penalty associated with CO_2 capture.

For example, sending gasification products of coal together with steam through a fluidized bed of lime would shift oxygen from water to carbon. Capture of CO_2 on lime would promote hydrogen production and provide necessary heat. Half of the hydrogen-rich output would be used to gasify coal; the other half would be oxidized in a high-temperature solid-oxide fuel cell. The water-rich spent fuel gas would be returned to the lime bed to repeat the cycle. Only excess water, ash, and impurities captured in various cleanup steps would leave the plant. Once the lime became fully carbonated limestone, CO_2 would be produced in a concentrated stream while the limestone was converted back to lime with waste heat from the fuel cell. Careful heat management could drive power plant efficiency to 70% (9) (for comparison, conventional coal-fired power plants are in the 30% to 35% range; modern gas-fired power plants can approach 50%).

CO_2 is three times as heavy as fuel and therefore cannot be stored in cars or airplanes. CO_2 from these sources will have to be released into the atmosphere and recaptured later. Currently, photosynthesis is the only practical form of air capture. Capture from air flowing over chemical sorbents—such as strong alkali solutions or activated carbon substrates—appears feasible but needs to be demonstrated (10). Wind is an efficient carrier of CO_2. The size of capture apparatus would be less than 1% that of windmills that displace equal CO_2 emissions, suggesting that they could be quite cheap to build (11). The additional cost of sorbent recycling should also be affordable (12).

Because the atmosphere mixes rapidly, extraction at any site, however remote, could compensate for emissions from anywhere else. By decoupling power generation from sequestration, air capture would allow the existing fossil fuel–based energy infrastructure to live out its useful life; it would open remote disposal sites and even allow for the eventual reduction of atmospheric CO_2 concentration.

Cost predictions for sequestration are uncertain, but $30 per ton of CO_2 (equivalent to $13 per barrel of oil or 25¢ per gallon of gas) appears achievable in the long term. Initially, niche markets (e.g., in enhanced oil recovery) would keep disposal costs low, with capture at retrofitted power plants dominating costs. Over time, new power plant designs could reduce capture costs, but the costs of disposal would rise as cheap sites filled up and demands on permanence and safety tightened. Some applications—for example, in vehicles and airplanes—could accommodate the higher price of CO_2 capture from air, eliminating CO_2 transport and opening up remote disposal sites.

Today's urgent need for substantive CO_2 emission reductions could be satisfied more cheaply by available sequestration technology than by an immediate transition to nuclear, wind, or solar energy. Further development of sequestration would assure plentiful, low-cost energy for the century, giving better alternatives ample time to mature.

References and Notes

1. H.-H. Rogner, *Annu. Rev. Energy Environ.* 22, 217 (1997).

2. M. I. Hoffert *et al.*, *Science* 298, 981 (2002).

3. The reason is that leakage rates are proportional to storage size and inversely proportional to storage lifetime.

4. K. S. Lackner, *Annu. Rev. Energy Environ.* 27, 193 (2002).

5. S. Holloway, *Annu. Rev. Energy Environ.* 26, 145 (2001).

6. H. Herzog, B. Eliasson, O. Kaarstad, *Sci. Am.* (February 2000), p. 72.

7. R. A. Chadwick *et al.*, paper presented at the Sixth International Conference on Greenhouse Gas Technology (GHGT-6), Kyoto, Japan, 1–4 October 2002.

8. H. J. Herzog, E. M. Drake, *Annu. Rev. Energy Environ.* 21, 145 (1996).

9. T. M. Yegulalp, K. S. Lackner, H.-J. Ziock, *Int. J. Surf. Mining Reclam. Environ.* 15, 52 (2001).

10. K. S. Lackner, H.-J. Ziock, P. Grimes, in *Proceedings of the 24th International Conference on Coal Utilization & Fuel Systems*, B. Sakkestad, Ed. (Coal Technology Association, Clearwater, FL, 1999), pp. 885–896.

11. At a wind speed of 6 m/s, the U.S. per-capita emission of 22 tons/year of CO_2 flows through an opening of 0.2 m². Through the same opening blow 21 W of wind power, or 0.2% of the U.S. per-capita primary power consumption.

12. F. S. Zeman, paper presented at the 2nd Annual Conference on Carbon Sequestration, Alexandria, VA, 5–8 May 2003.

Carbon Trading

WILLIAM H. SCHLESINGER

Enthusiasm is spreading for cap and trade systems to regulate the amount of carbon dioxide (CO_2) emitted to Earth's atmosphere. In 1990, the U.S. Environmental Protection Agency set a limit on sulfur dioxide (SO_2) emissions from obvious point sources and allowed those who emit less than their quota to trade excess allowances. As a result, regional acid deposition was dramatically reduced. Can the world do the same for CO_2?

Fundamental differences in the biogeochemistry of SO_2 and CO_2 suggest that establishing a comprehensive, market-based cap and trade system for CO_2 will be difficult. For SO_2, anthropogenic point sources (largely coal-fired power plants), which are relatively easy to control, dominate emissions to the atmosphere.

Natural sources, such as volcanic emanations, are comparatively small, so reductions of the anthropogenic component can potentially have a great impact, and chemical reactions ensure a short lifetime of SO_2 in the atmosphere. CO_2, in contrast, comes from many distributed sources, some sensitive to climate, others sensitive to human disturbance such as cutting forests. It is thus impossible to control all of the potential sources.

Human-derived emissions from fossil fuel

This article first appeared in *Science* (24 November 2006: Vol. 314, no. 1217). It has been revised for this edition.

combustion are one of the smaller components of the atmospheric flux of CO_2, which is dominated by exchange between forests and the oceans. During most of the past 10,000 years, the uptake and loss of CO_2 from forests and the oceans must have been closely balanced, because atmospheric CO_2 showed little variation until the start of the Industrial Revolution. CO_2 from coal, oil, and natural gas combustion now comes from many segments of society, including electric power generation, industry, home heating, and transportation. Unbalanced by equivalent anthropogenic sinks for carbon, fossil fuel emissions account for the vast majority of the rise of CO_2 in Earth's atmosphere. Caps on emissions, like those instituted for SO_2, will be difficult to institute if the burden of reducing CO_2 is to be borne equally by all emitters.

Because land plants take up CO_2 in photosynthesis and store the carbon in biomass, forests and soils seem to be attractive venues to store CO_2. Market-based schemes propose substantial payments and credits to those who achieve net carbon storage in forestry and agriculture, but these projected gains are often small and dispersed over large areas. We will need to net any such carbon uptake against what might have occurred without climate-policy intervention. Conversely, will Canada and Russia be billed for incremental CO_2 releases that stem from the warming of cold northern soils as a result of global warming from the use of fossil fuels worldwide?

If credit is given to those who choose not to cut existing forests, the increasing total demand for forest products will shift deforestation to other areas. Frequent audits will be needed to determine current carbon uptake, insurance will be necessary to protect past carbon credits from destruction by fire or windstorms, and payments will be necessary if the forest is cut. All these efforts will be costly to administer, diminishing the value of the rather modest carbon credits expected from forestry and agriculture.

Many environmental economists recognize that a tax or fee on CO_2 emission from fossil fuel sources is the most efficient system to reduce emissions and spread the burden equitably across all sources: industrial and personal. A tax on emissions of fossil fuel carbon could replace the equivalent revenue from income taxes, so the total tax bill of consumers would be unchanged. A higher tax on gasoline would preserve the personal right to drive a larger car or drive long distances, but it would also motivate decisions to do otherwise. A tax on emissions from coal-fired power plants, manifest in monthly electric bills, would motivate the use of alternative energies and energy-use efficiencies at home and in industry.

The biogeochemistry of carbon suggests that both emissions taxes and cap and trade programs will work best if restricted to sources of fossil fuel carbon. Other net sources and sinks of carbon in its global biogeochemical cycle are simply too numerous and usually too small to include in an efficient trading system. *Simple, fair,* and *effective* must be the hallmarks of policies that will wean us from the carbon-rich diet of the Industrial Revolution, and we must begin soon if we are to have any hope of stabilizing our climate.

Carbon Trading Over Taxes

WILLIAM CHAMEIDES AND MICHAEL OPPENHEIMER

A s the United States moves inevitably toward climate legislation, discussion has shifted from the science to the policy options for slowing emissions of carbon dioxide (CO_2) and other greenhouse gases. Some favor a tax on CO_2 emissions—referred to as a C tax (1). Others favor government subsidies (2). If high enough to alter consumer behavior, a carbon tax would reduce emissions by raising the effective price of carbon-intensive energy relative to carbon-free sources. Subsidies may speed development of specific, targeted low-C technologies.

But a market-based system with an economywide cap on emissions and trading of emission allowances would do the same, while having distinct advantages (3). Most important, a

---

This article first appeared in *Science* (23 March 2007: Vol. 315. no. 5819). It has been revised for this edition.

cap and trade system, coupled with adequate enforcement, would assure that environmental goals actually would be achieved by a certain date. Given the potential for escalating damages and the urgent need to meet specific emission targets (4), such certainty is a major advantage. A federal cap and trade system could be incorporated into existing emissions-trading frameworks and markets, such as the Kyoto Protocol's international market or subnational ones like the Regional Greenhouse Gas Initiative.

Earth's climate is agnostic about the location and type of CO_2 emissions and is sensitive only to the total burden of CO_2. It makes sense, therefore, to design a climate policy that taps all possible avenues to limit net CO_2 emissions. Trading of emissions across all sectors of the economy addresses this by allowing emitters to purchase carbon offsets from businesses that are able to lower their own emissions below their allocation. If trading were incorporated into an international system, U.S. firms and consumers could meet emissions targets at reduced costs by substituting less expensive cuts in, for example, developing countries, for expensive emissions cuts in the United States. Because investment would be funneled to technologies that reduce CO_2 emissions at the least cost, the overall expense of the program would be minimized.

Cutting emissions of pollutants is admittedly not as complicated as cutting CO_2 emissions, and transaction costs can be a factor. Nevertheless, the United States was able to reduce sulfur oxide emissions ahead of schedule and at 30% of the projected cost using a market-based cap and trade system (5). Elimination of lead from gasoline and phaseout of ozone-depleting chemicals were also facilitated by emissions-trading programs.

Offsetting emissions by storing carbon in soils, forests, and other forms of biomass in the United States has the potential to offset 10% to 20% of U.S. emissions in 2025 at relatively low cost (see table S1 online and Image 1). International opportunities also exist. Deforestation of tropical rainforests is currently estimated to cause more than 7000 million metric tons per year of CO_2 emissions, the equivalent of about 25% of worldwide emissions from fossil fuel burning today; in 2025 the percentage is estimated to be about 15% (see table S2 online). Using an international cap and trade market to compensate nations for slowing deforestation would bring a significant block of emissions under management, while preserving irreplaceable ecosystems and providing income to developing economies (6–8).

Ensuring the integrity of such a system will require rigorous monitoring, auditing, and registration. Leakage (e.g., where reduced timber harvest in one location is replaced by increased harvest elsewhere to meet demand for lumber), the credibility of baselines in capped and uncapped systems, and the full climate effects of enhanced biological growth must be addressed (9–10). However, these problems

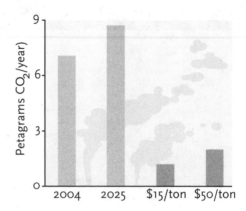

IMAGE 1 Effect of carbon credits. U.S. greenhouse gas emissions for 2004 and 2025 with business as usual (blue); estimated U.S. land management offsets in 2025 at \$15 and \$50 per ton of CO_2 (green); 1 petagram = 1 billion tons (table S1).

are manageable (11). Frameworks and methodologies for documenting the size and validity of carbon offsets based on land management practices are available (12, 13). Following such a methodology will not be a trivial exercise. It will involve costs that will affect those hoping to market offsets. But the advantage of a market-based system is that it provides an incentive for innovation—which can translate into inexpensive CO_2 emission reductions. Why would we want to exclude any sector of the economy from this competition, let alone one with such large potential?

References and Notes

1. W. H. Schlesinger, *Science* 314, 1217 (2006).
2. M. I. Hoffert *et al.*, *Science* 298, 981 (2002).
3. C.-J. Yang, M. Oppenheimer, *Clim. Change* 80, 199 (2007).
4. H. J. Schellnhuber, Ed., *Avoiding Dangerous Climate Change* (Cambridge Univ. Press, Cambridge, 2006).
5. *National Acid Precipitation Assessment Program Report to Congress: An Integrated Assessment* (U.S. National Science and Technology Council, Washington, DC, 2005); www.esrl.noaa.gov/csd/AQRS/reports/napapreport05.pdf.

*Science*NOW Daily News, 6 December 2007

SENATE PANEL ADOPTS EMISSIONS CURBS
Eli Kintisch

A bill that would cut U.S. greenhouse gas pollution by 70% in 2050, relative to 2005 levels, was approved by a Senate panel last night in what supporters are hailing as a landmark vote in the fight to mitigate global warming. "Today, the Senate took a giant and historic step forward toward reversing a clear and present danger to our planet," said Senator Joseph Lieberman (I–CT) in a statement issued last night.

Yesterday's vote came after dozens of hearings by the Senate Environment and Public Works committee since Democrats took control of Congress in January. Known as America's Climate Security Act of 2007 (S.2191), the bill would create a system in which businesses that create or deal with carbon emissions would be issued or sold allowances. This "cap and trade" system would allow them to emit greenhouse gases up to that level or trade the allowances if they could otherwise reduce pollution from operations with clean-energy technology. Introduced by Lieberman and Republican lawmaker John Warner of Virginia, who cast the only Republican vote in favor of the 303-page measure, the legislation affects everything from power plants to forests to elderly consumers facing rising electricity prices.

Yesterday's debate foreshadowed a number of hurdles that could prevent the legislation from ever becoming law. Republicans unsuccessfully offered amendments that would have automatically shuttered the system if more than 10,000 automaker jobs were lost or if experts found that it was not reducing world temperatures effectively. These failed on largely party-line votes, with panel chair Barbara Boxer (D–CA) repeatedly emphasizing that such "poison bill" amendments could upend the fragile coalition of environmental groups and selected industries that support the bill.

The Senate is unlikely to take any further action on the bill until next year, and getting the 60 votes needed to avoid an expected filibuster won't be easy. Beyond Warner, only a handful of Republicans are sympathetic to the bill, including senators Susan Collins (R–ME) and John McCain (R–AZ). But McCain demands more support for nuclear energy, an issue on which Democrats won't easily budge. Previewing what is expected to be a massive campaign against the bill, the U.S. Chamber of Commerce recently launched advertising on the Internet and in Washington, D.C.–area airports. The ads, featuring suburban professionals cooking with candles and jogging to work in suits, argue that the legislation would make energy too expensive.

Despite the remaining challenges, supporters see the committee vote as a victory. "Even if this bill doesn't pass the Senate and House next year, it is likely to be the blueprint for action early in the next president's term," said the Pew Environment Group in a statement.

6. P. Moutinho, S. Schwartzman, Eds., *Tropical Deforestation and Climate Change* (Amazon Institute for Environmental Research, Belém, Pará, Brazil, 2005).

7. R. Bonnie, S. Schwartzman, M. Oppenheimer, J. Bloomfield, *Science* 288, 1763 (2000).

8. M. Santilli *et al.*, *Clim. Change* 71, 267 (2005).

9. F. Keppler *et al.*, *Nature* 439, 187 (2006).

10. S. Gibbard *et al.*, *Geophys. Res. Lett.* 32, L23705 (2005).

11. L. Olander, *Do Recent Scientific Findings Undermine the Climate Benefits of Sequestration in Forests?* (Nicholas Institute, Durham, NC, 2006); www.nicholas.duke.edu/institute/methanewater.pdf.

12. Nicholas Institute, *Harnessing Farms and Forests in the Low-Carbon Economy: How to Create, Measure,* and Verify Greenhouse Gas Offsets Based on Storing Carbon in Trees and Soil and Reducing Emissions from Land Use (Duke Univ. Press, Durham, NC, in press).

13. J. M. Penman *et al.*, Eds., *Intergovernmental Panel on Climate Change: Good Practice Guidance for Land Use, Land-Use Change, and Forestry* (Institute for Global Environmental Strategies, Hayama, Kanagawa, Japan, 2003); www.ipcc-nggip.iges.or.jp/public/gpglu lucf/gpglulucf.htm.

Supporting Online Material

www.sciencemag.org/cgi/content/full/315/5819/ 1670/DC1

Index